THE NATIONAL SPACE TRANSPORTATION POLICY

Issues for Congress

Office of
Technology
Assessment

Congress
of the
United States

Recommended Citation: U.S. Congress, Office of Technology Assessment, *The National Space Transportation Policy: Issues for Congress,* OTA-ISS-620 (Washington, DC: U.S. Government Printing Office, May 1995).

For sale by the U.S. Government Printing Office
Superintendent of Documents, Mail Stop: SSOP, Washington, DC 20402-9328
ISBN 0-16-048077-9

Foreword

In responding to the political and military challenges of the Cold War, and the urge to explore and exploit outer space, the United States developed a capable fleet of space transportation systems for carrying cargo and people into space, and for ensuring a credible strategic nuclear deterrent. These systems are owned and managed by the National Aeronautics and Space Administration, the Department of Defense, and private industry. In recent years, increasing federal budget constraints, commercial competition from foreign launch firms, and a desire to continue an ambitious space program have created pressures within the United States to reduce the costs of access to space. Significantly lower space transportation costs would make the U.S. space industry more commercially competitive, foster the expansion and creation of new space markets, and ensure access to space for government payloads and manned missions.

This report, prepared for the House Committee on Science, is the first in a broad assessment of the health and future prospects of the U.S. space transportation technology and industrial base. The report focuses on the Clinton Administration's National Space Transportation Policy, which was released last fall. It examines administration policy in light of the implementation plans prepared by NASA, DOD, and the Transportation and Commerce Departments. As the report notes, the new policy brings a welcome measure of order to the sometimes chaotic structure of U.S. space transportation activities. The policy also emphasizes the important contribution private industry can make to the direction and development of U.S. space transportation capabilities. However, an analysis of the policy and implementation plans also raises some issues that might be of interest to Congress as it debates space transportation legislation, oversight, and funding. These issues involve decisions on NASA and DOD development programs, the use of foreign launch vehicles and components, the conversion of excess long-range ballistic missiles for use as launch vehicles, and the new role of the private sector in space transportation research and development decisionmaking. This report also identifies two issues omitted from the Administration's policy: the preservation of long-range ballistic missile capabilities after final production in 2005, and the perspective of lower industrial tier firms toward national space transportation policy.

In undertaking this effort, the Office of Technology Assessment sought the contributions of a wide spectrum of knowledgeable individuals and organizations. Some provided information, others reviewed drafts. OTA gratefully acknowledges their contributions of time and intellectual effort. OTA also appreciates the help of NASA and the Defense, Transportation, and Commerce Departments. As with all OTA reports, the content of this report is the sole responsibility of OTA and does not necessarily represent the views of our advisors or reviewers.

ROGER C. HERDMAN
Director

Advisory Panel

Ronald Brunner, *Chairman*
Director
Center for Public Policy Research
University of Colorado

Edward C. Aldridge, Jr.
President and CEO
The Aerospace Corporation

Buzz Aldrin
Apollo 11 Astronaut and Consultant

Radford Byerly, Jr.
Consultant

Thomas Burson
Vice President and General Manager, Space Transportation
McDonnell Douglas Aerospace

Paul J. Coleman, Jr.
Director
National Institute for Global Environmental Change
University of California at Los Angeles

Lt. Gen. Donald L. Cromer, USAF (retired)
President
Hughes Space and Communications Co.

Henry J. Dinenno
Vice President, Advanced Programs & Business Development
Space Systems Division
Rockwell International Corp.

Isaac T. Gillam, IV
Senior Vice President
OAO Corporation

Michael D. Griffin
Senior Vice President, Program Development
Space Industries

Frederick H. Hauck
President and CEO
INTEC

Clark W. Hawk
Director
Propulsion Research Center
University of Alabama at Huntsville

Douglas A. Heydon
President
Arianespace, Inc.

Joan Johnson-Freese
Associate Professor
Department of National Security Studies
Air War College

Jon B. Kutler
President
Quarterdeck Investment Partners, Inc.

Ronald G. Peterson
Vice President and General Manager, Space/Strategic Propulsion
Hercules Aerospace Co.

James D. Phillips
Director of Engineering Development (Retired)
Kennedy Space Center

Thomas F. Rogers
President
Sophron Foundation

Jerome Simonoff
Consultant

Larry N. Speight
Vice President, Space and Strategic Systems
Honeywell

Courtney A. Stadd
Managing Partner
Global Technology Ventures

Peter B. Teets
President, Information and Services Sector
Lockheed Martin Corp.

David W. Thompson
President and CEO
Orbital Sciences Corp.

Joseph P. Zimonis
Executive Vice President and General Manager
USBI Co.

Note: OTA appreciates and is grateful for the valuable assistance and thoughtful critiques provided by the advisory panel members. The panel does not, however, necessarily approve, disapprove, or endorse this report. OTA assumes full responsibility for the report and the accuracy of its contents.

Project Staff

Peter Blair
Assistant Director
Industry, Commerce, and International Security Division

Alan Shaw
Program Director
International Security and Space Program

PROJECT STAFF

Christopher M. Waychoff
Project Director

S. Yousef Hashimi
Analyst

Jack Nunn
Senior Associate

Mark Suskin
Analyst

Ray A. Williamson
Senior Associate

ADMINISTRATIVE STAFF

Jacqueline Robinson Boykin
Office Administrator

N. Ellis Lewis
Administrative Secretary

Don Gallagher
Secretary

CONTRACTORS

Leonard David
Donald Fowler
Carol Gaelick
Joel S. Greenberg
Ivars Gutmanis
Elizabeth Sheley

INTERNAL REVIEWERS

Kevin Dopart
Thomas H. Karas
Peter G. Smith
Michael G. Snyder

Acknowledgments

Jonathan C. Ball
U.S. Department of Commerce
Washington, DC

Col. Charles R. Banta
U.S. Air Force
Washington, DC

Mark Bitterman
Orbital Sciences Corp.
Dulles, VA

C.R. Carlson
McDonnell Douglas Aerospace
Huntington Beach, CA

JoAnn Clayton
National Research Council
Washington, DC

Lee Edwards
U.S. General Accounting Office
Huntsville, AL

Don Eiss
Office of the U.S. Trade Representative
Washington, DC

William A. Gaubatz
McDonnell Douglas Aerospace
Huntington Beach, CA

Dennis Granato
U.S. Deparment of Defense
Washington, DC

Jaak Holemans
Booz-Allen & Hamilton, Inc.
Arlington, VA

Dave Hubbell
U.S. General Accounting Office
Los Angeles, CA

Alan Ladwig
National Aeronautics and Space Administration
Washington, DC

J. Wayne Littles
National Aeronautics and Space Administration
Washington, DC

Capt. Gavin D. Lowder
U.S. Navy
Washington, DC

John E. Mansfield
National Aeronautics and Space Administration
Washington, DC

Larry Mattson
TRW
Redondo Beach, CA

Wallace A. McClure
Westminster, CA

Richard L. McMillion
Rocketdyne
Canoga Park, CA

Henry M. Minami
Rocketdyne
Canoga Park, CA

David H. Moore
U.S. Congressional Budget Office
Washington, DC

Warren G. Morimoto
Pratt and Whitney
West Palm Beach, FL

Arthur C. Morrissey
Lockheed Martin Corp.
Crystal City, VA

David E. Mosher
U.S. Congressional Budget Office
Washington, DC

Eric E. Nichols
Naval Space Command
Dahlgren, VA

William Piland
NASA Langley Research Center
Hampton, VA

Karen S. Poniatowski
National Aeronautics and Space Administration
Washington, DC

Col. Charles Pugsley
U.S. Air Force
Arlington, VA

David P. Radzanowski
U.S. Congressional Research Service
Washington, DC

Jim Sanders
NASA Marshall Space Flight Center (retired)
Huntsville, AL

Richard Scott
U.S. Department of Transportation
Washington, DC

Lt. Col. Jess Sponable
USAF Phillips Laboratory
Albuquerque, NM

Homer Thompson
U.S. General Accounting Office
Washington, DC

Sam Van Wagner
U.S. General Accounting Office
Los Angeles, CA

Note: OTA appreciates and is grateful for the valuable assistance and thoughtful critiques provided by these individuals. They do not, however, necessarily approve, disapprove, or endorse this report. OTA assumes full responsibility for the report and the accuracy of its contents.

Acronyms and Initialisms

BMDO	Ballistic Missile Defense Organization
CAN	cooperative agreement notice
CSTS	Commercial Space Transportation Study
DOC	Department of Commerce
DOD	Department of Defense
DOT	Department of Transportation
EELV	Evolved Expendable Launch Vehicle
ELV	expendable launch vehicle
ESA	European Space Agency
FSU	former Soviet Union
GEO	geosynchronous Earth orbit
GTO	geosynchronous transfer orbit
HLV	heavy launch vehicle
ICBM	intercontinental ballistic missile
LFBB	Liquid Fly Back Booster
LLV	Lockheed Launch Vehicle
MLV	medium launch vehicle
MTCR	Missile Technology Control Regime
NASA	National Aeronautics and Space Administration
NASP	National Aerospace Plane
nmi	nautical miles
NSTC	National Science and Technology Council
NSTP	National Space Transportation Policy
OSTP	Office of Science and Technology Policy
OTA	Office of Technology Assessment
R&D	research and development
RFP	request for proposal
RLV	reusable launch vehicle
SLBM	submarine-launched ballistic missile
SLV	small launch vehicle
SRB	solid rocket booster
SSTO	single-stage-to-orbit (vehicle)
START	Strategic Arms Reduction Talks
TSTO	two-stages-to-orbit (vehicle)
USRA	Universities Space Research Association
USTR	U.S. Trade Representative

Contents

Summary

As the year 2001 approaches, visions of annular space stations and tourist flights to the Moon remain science fiction. With fewer than six years until the new millennium, the U.S. space transportation technology and industrial base is faced with a number of challenges, and some opportunities.[1]

The federal government has been the primary customer for space transportation services since the early days of rocketry. Recent efforts to reduce the federal budget deficit, cut the national debt, and shift development responsibilities to the private sector, however, have limited government funding for new space transportation technologies and missions. Similarly, the end of the Cold War has led to a reexamination of defense and national security spending on space missions and long-range ballistic missiles.

Meanwhile, European, Russian, and Chinese launch providers have captured more than 60 percent of the global commercial launch market for medium launch vehicles (MLVs). At the same time, entrepreneurs in the telecommunication, navigational, and remote sensing satellite industries predict an increasing need for launch services to establish and maintain large constellations of new satellites.

[1] Space transportation in this report refers to vehicles able to carry payloads or passengers to orbit. Space transportation systems may be expendable launch vehicles (ELVs), partially or fully reusable launch vehicles (RLVs), and long-range ballistic missiles. Currently, the U.S. Space Shuttle is the world's only operational RLV. This report does not address suborbital launch systems or transportation systems designed primarily to move payload or passengers between or beyond Earth orbits.

To address these challenges and opportunities, on August 5, 1994, the White House announced the National Space Transportation Policy (NSTP), developed by the National Science and Technology Council and approved by President Clinton.[2] The Clinton Administration's four fundamental objectives for the NSTP were to establish new national policy regarding:

1. federal space transportation spending, consistent with current budget constraints and the opportunities presented by emerging technologies;
2. federal agencies' use of foreign launch systems and components;
3. federal agencies' use of excess ballistic missile assets for space launch, to prevent adverse impacts on the U.S. commercial space launch industry; and
4. an expanded private sector role in the federal space transportation research and development (R&D) decisionmaking process.[3]

This report examines the new policy and the implementation plans of the Department of Defense (DOD), the National Aeronautics and Space Administration (NASA), and the Departments of Transportation (DOT) and Commerce (DOC) in the context of these four fundamental objectives. The report raises issues for Congress to consider as it debates the funding, oversight, and legislative requirements of the new space transportation policy. General findings, issues for Congress, and critical decision points identified in the report are summarized below. The main body of the report provides background on the fundamental objectives and examines each issue for Congress in detail.

GENERAL FINDINGS

Lack of Consensus on U.S. Space Policy Goals

The U.S. space transportation technology and industrial base is in a period of tumult and uncertainty brought about by the end of the Cold War, a constrained fiscal environment, and a pending shift in responsibilities from the public to the private sector. Even more than ordinary times, such a period demands clear intermediate and long-term goals, strong Presidential leadership, and the formation of a national consensus among the executive branch, Congress, industry, and the public.

The NSTP states that "assuring reliable and affordable access to space through U.S. space transportation capabilities is a fundamental goal of the U.S. space program."[4] Most observers agree that reducing costs and improving reliability are important objectives for the U.S. space program. Reliable, affordable access to space is a necessary part of the nation's infrastructure for achieving broader space goals.

The policy implies that lowering the cost of access to space will allow the United States to do whatever it wants in space. It may be difficult to achieve lower launch costs, however, without a clearly articulated, long-term plan supported by adequate funding, especially when the government is asking industry to make significant investments in ambitious new space transportation development programs. Indeed, the Office of Technology Assessment (OTA) previously noted that "until the nation chooses *what* it wants to accomplish in space, and what the U.S. taxpayer is

---

[2] Presidential Decision Directive NSTC-4. Most, if not all, the text of this internal policy was released publicly in The White House, Office of Science and Technology Policy, "National Space Transportation Policy," Fact Sheet, Washington, DC, Aug. 5, 1994. See appendix for complete text.

[3] The White House, Office of Science and Technology Policy, "Statement on National Space Transportation Policy," Washington, DC, Aug. 5, 1994. See appendix for complete text.

[4] The White House Office, Office of Science and Technology Policy, op. cit., footnote 2, Intro.

willing to pay for, neither the *type* nor number of necessary launchers and facilities can be estimated with accuracy."[5]

Establishing a national consensus first requires a clear delineation of specific national space goals by the Administration and its implementing departments and agencies. Then the Administration must cultivate congressional and public support for these goals, and convince industry that pursuing and achieving these goals would serve its interests.

The Administration has outlined some broad national space goals, such as achieving the International Space Station. It has not made clear, however, how specific goals relate to each other. It has not issued, for example, an overall space policy to replace the 1989 space policy of the Bush Administration. Without a clear articulation of space policy and how it relates to the broader national agenda, it may be difficult for both industry and government to pursue space transportation goals with vigor.

As the experience of the last decade has shown, even if the Administration were to delineate clear and specific national space goals, industry officials might still be reluctant to commit corporate resources to new space transportation ventures without strong congressional support. This support, in turn, depends on the ability of Members of Congress to bridge jurisdictional divisions and reach a consensus on how to buttress national space goals.

Living Within Severe Budget Constraints

Fiscal constraints imposed by the budget deficit, the federal debt, competition from other programs, and a desire to reduce government spending have forced DOD and NASA to cut back on space transportation programs, and to attempt to fund more creatively those programs that remain (e.g., through public-private partnerships). Both DOD and NASA state that they can meet their current space goals, but many government and industry officials are skeptical. These officials point out the previous commitments to new space transportation systems that failed to produce operational vehicles despite less severe budget constraints (e.g., the National Launch System, the Advanced Launch System, the Air Force Space Lifter, the Shuttle C, the Shuttle II, and the National Aerospace Plane). The U.S. government could afford to fund fully new space transportation systems, but it has currently placed its spending priorities elsewhere. In the absence of more government spending, the government and industry will have to sustain a commitment to new ways of doing business if the challenges and opportunities confronting U.S. space transportation are to be met.

Government Demand Dominates the Space Transportation Market

Since the advent of the space age, the U.S. government has played a large and critical role in shaping the domestic space transportation technology and industrial base. The U.S. government created the base to build long-range ballistic missiles and place men on the Moon. The U.S. government and industry remain tightly entwined through R&D and procurement contracts, federal regulations, and the need for licenses, despite the rise of commercial space launch markets.

For at least the next decade, U.S. national security and civil demands for space transportation will continue to dominate the domestic industry. Even the most optimistic growth projections for the global commercial market do not forecast a significant shift away from the government without major changes in the marketplace (e.g., the development of a dramatically less-expensive space transportation system or the discovery of a commercially lucrative space activity). Moreover, some launch providers are reluctant to take the steps necessary to make their launch operations

[5] U.S. Congress, Office of Technology Assessment, *Access to Space: The Future of U.S. Space Transportation Systems*, OTA-ISC-415 (Washington, DC: U.S. Government Printing Office, April 1990), p. 21.

more commercially price-competitive, because the changes might conflict with government requirements or the government might demand similar savings.

Current Capabilities Can Meet National Security Requirements

The national security community currently requires a domestic capability to launch payloads into orbit. The existing fleet of launch vehicles can continue to meet this requirement for the foreseeable future. DOD's major new development program, the Evolved Expendable Launch Vehicle (EELV), is intended to reduce DOD space transportation costs, rather than ensure continued access to space. The EELV program attempts to maximize cost savings by replacing DOD's current stable of MLVs and heavy launch vehicles (HLVs), procured from several vendors, with a unified family of vehicles that share many common components and launch infrastructure and are built by a single launch provider. One consequence of this consolidation, however, is that a systemic failure in one vehicle might ground the entire family of vchicles along with its national security payloads.

Competing Interests Make Common Strategy Difficult

While all members of the space transportation technology and industrial base see a critical need to reduce the cost of space transportation, their differing interests make it difficult to agree on a common space transportation development strategy. National security space transportation decisions are largely driven by the need to reduce expensive HLV launch costs through the EELV program. NASA hopes to replace its aging Space Shuttle fleet with a new, low-cost reusable launch vehicle (RLV) in the high MLV class that would carry crews and cargo to and from the International Space Station. Most industry representatives want to focus development dollars on a smaller, reusable MLV, designed to recapture lost commercial market share worldwide, while small launch vehicle (SLV) producers and selected space scientists want to maintain U.S. leadership in SLVs.

Recognizing and balancing these competing interests is critical to the success of a truly national space policy. The NSTP and its implementation plans are careful to ensure that DOD and NASA needs are met, but are less diligent about meeting the needs of the private sector.

ISSUES FOR CONGRESS

The following section summarizes OTA's discussion of issues that may be of interest to Congress as it debates the future of U.S. space transportation. These issues are divided among the Clinton Administration's four fundamental objectives for the NSTP and correspond directly to the main body of the report.

Fundamental Objective #1: Space Transportation Funding and the Division of Responsibilities

The NSTP attempts to set government spending priorities for current and future space transportation systems by assigning specific roles and functions to designated departments and agencies. By placing DOD in charge of expendable launch vehicle (ELV) development and NASA in charge of continued Space Shuttle operation and RLV development, the Administration has taken a step toward reducing conflicts and redundancies within government space transportation development programs.

DOD currently spends roughly $1.6 billion (about 84 percent) of its $1.9-billion space transportation budget on its HLV program, the Titan IV, while NASA spends just over $4 billion per year on Space Shuttle modifications and operations. Each organization has individually initiated a set of programs to address the budgetary difficulties posed by these high costs (see table S-1).

An added dimension to the current effort to develop new space transportation systems is the role of the private sector—both in decisionmaking and

TABLE S-1: DOD and NASA Development Programs and Procurements

	Program	Description	Objective
DOD	Existing ELV upgrades	Upgrades to the current fleet of launch systems and supporting infrastructure.	Keep the existing ELV fleet flying safely and reliably. Achieve significant, short-term payoffs where possible.
	Evolved Expendable Launch Vehicle (EELV)	A new, single family of MLVs and HLVs based on an evolutionary redesign of one or more existing ELVs.	Lower overall cost of access to space for DOD, especially for heavy payloads, by using common subsystems, components, and infrastructure.
NASA	Space Shuttle upgrades	System improvements and replacement of aging subsystems and components.	Keep the Space Shuttle flying safely and reliably.
	Med-Lite	A launch system with capacity falling between existing SLVs and MLVs.	Acquire a less expensive vehicle to serve future planetary exploration mission requirements.
	DC-XA	Upgrades to the DC-X (a vertical-takeoff/vertical-landing, sub-scale, fully reusable, single-stage-to-orbit technology demonstrator) with more advanced components.	Demonstrate system operability by testing new components and the integrated system in a realistic flight environment.
	X-33	A sub-scale advanced technology demonstrator that will be, at a minimum, an autonomous, suborbital, experimental, single-stage-to-orbit RLV.	By 2000, prove the concept of a fully reusable single-stage-to-orbit space transportation system in the high MLV class by demonstrating key technology, operations, and reliability in an integrated flight vehicle. Encourage private investment in a commercial follow-on RLV by reducing the technical risks of SSTO.
	X-34	A partially reusable demonstration vehicle for small payloads.	Investigate technologies that may be incorporated into future RLVs. Demonstrate streamlined management of joint industry-government development effort. Address commercial and U.S. government need for an inexpensive SLV.
	Supporting RLV technology demonstration	Development and validation of propulsion, structural, and operations technologies.	Progressively integrate and flight-demonstrate these technologies on three experimental test vehicles (DC-XA, X-33, and X-34) in order to 1) mature technologies required for a next-generation launch system, 2) demonstrate the capability to achieve low development and operational costs, and rapid launch turnarounds, and 3) reduce technical risk to encourage private investment in the commercial development and operation of next-generation systems.

SOURCE: Office of Technology Assessment, 1995.

financing. This has important implications on the nature of space transportation development programs and raises an entirely new set of considerations that must be taken into account when evaluating development programs proposed by both DOD and NASA.

ISSUE 1a: *Divided responsibility and interagency coordination*

The NSTP divides the government's primary responsibilities for space transportation between DOD and NASA. If existing space transportation assets and those under development are to be managed in a manner conducive to all interests, this division of responsibility will increase the need for both organizations to coordinate with one another, as well as with the private sector.

That DOD and NASA will adequately account for the interests of all parties is not a certainty, especially as funds available for space transportation diminish. Conflicts over how to approach the development of new space transportation systems will undoubtedly arise. At present, it appears that resolution of these conflicts will be achieved via negotiations between DOD and NASA on a case-by-case basis, possibly with some mediation by a third party within the executive branch. Such negotiations may succeed in satisfying both DOD and NASA, but could fail to account for the interests of all relevant parties, especially those in the private sector.

Such negotiations could also lead to programmatic redundancies. In the absence of central authority or leadership, DOD and NASA may discount potential redundancies and simply continue to promote those projects that best address their own organizational requirements. As a result, hard space transportation policy choices may go unmade.

Many analysts and policymakers have proposed a central authority as a way to better account for all interests and avoid programmatic redundancies. In the Bush Administration, for example, Vice President Quayle was given considerable authority over space transportation policy. It is not clear, however, that the imposition of a central authority has remedied these problems in the past, or that it will necessarily do so in the future. Given the considerable bureaucratic and political weight of DOD and NASA, competing organizational interests could potentially override the wishes of a central authority. Furthermore, existing legal and organizational obstacles may prevent the level of interagency and private sector coordination sought by a central authority.

The recent controversy between DOD and NASA over NASA's Med-Lite procurement may be emblematic of this latter problem. This controversy illustrates how interagency coordination can be precluded by current law, divergent interpretations of that law, and competing organizational interests. Therefore, although the NSTP calls on DOD and NASA to "combine their [ELV] requirements into single procurements when such procurements would result in cost savings or are otherwise advantageous to the government," achieving this level of interagency coordination may prove extremely difficult.

ISSUE 1b: *Potential conflicts and redundancies*

DOD and NASA have collectively proposed a sizable portfolio of new space transportation technology development programs. While this multitrack approach may reduce the overall risk of pursuing new space transportation systems, it may also lead to potential conflicts and redundancies and higher overall costs. For example, development of a commercially competitive EELV by DOD could undercut NASA's effort to commercialize a follow-on to the X-33 by reducing the incentive for private investors to fund a technically risky RLV.

If a low-cost RLV is developed, nonetheless, it may be difficult for the EELV program to achieve the long-term cost reduction targets set by DOD. At a minimum, the RLV will compete with the EELV for payloads, possibly attracting payloads away from the EELV. Were this to occur, it would reduce savings generated by the EELV program by offsetting or potentially outweighing any gains in production volume created by commonality within the EELV family.

As for NASA's dual-track RLV development strategy, the Agency believes that early X-34 test flights could positively affect X-33 development by steering it toward or away from certain technologies. Moreover, proponents note that the X-34 could generate significant benefits for the government, industry, and consumers of space-based services if its target of threefold cost reductions for launching small payloads are achieved.

Critics, however, have suggested canceling the X-34 program, arguing that it is geared toward developing an operational vehicle, not an experimental vehicle, and that its cancellation would not affect the technological success of the X-33 program.

There are other potential conflicts and redundancies. In particular, DOD officials are concerned that the Med-Lite vehicle might eventually compete with the EELV for medium payloads, thereby threatening the ability of the EELV program to achieve maximum launch cost reductions for DOD. In addition, NASA has proposed to phase in any newly developed RLV follow-on to the X-33 between 2005 and 2012 while continuing to fly the Space Shuttle in support of the International Space Station.

ISSUE 1c: *HLVs drive the EELV program*

DOD currently spends $1.6 billion of its $1.9-billion space transportation budget on its HLV program. Therefore, DOD has geared the EELV program toward achieving significant HLV cost reductions. DOD's focus on HLV cost reductions, however, ignores the need of U.S. launch providers to develop a commercially competitive launch vehicle in the hotly contested MLV market. While the EELV program may reduce MLV costs by as much as 10 percent, such a cost reduction would probably not help the EELV manufacturer recapture a significant portion of the global market for launch services. And without an increased share of the available market, DOD will receive little, if any, additional price reductions generated by larger production volumes.

On the other hand, the European Space Agency's (ESA's) development of the heavy-lift Ariane 5 suggests that significant HLV cost reductions may be commercially attractive. It is unlikely, however, that the heavy-lift EELV will be inexpensive enough to compete with the Ariane 5, even if it achieves a 40 percent cost reduction over the Titan IV.

ISSUE 1d: *RLV development*

NASA has pointed to its RLV development strategy as one example of its "new way of doing business." Outside of NASA, however, some have expressed concern over NASA's proposed RLV development strategy. In particular, industry officials are concerned that property and data rights issues, as well as the uncertainty surrounding the distribution of core RLV technology development funds, may slow or prevent RLV commercialization.

Other analysts and policymakers believe the X-33 program should be structured differently. Some critics have argued that the X-33 is being designed in the shadow cast by future requirements for the International Space Station, and that it would be better if NASA opted to fund fully an X-33 program that focuses solely on demonstrating single-stage-to-orbit (SSTO) technology. NASA officials believe the X-33 does focus on demonstrating SSTO technology, but contend that industry investment is appropriate because the successful development of a low-cost, commercial RLV will significantly improve the competitiveness of the U.S. space transportation industry, and because government space budgets are declining.

Others critics of the X-33 program structure have suggested a competitive fly-off among competing X-33 concepts, a strategy in which NASA has expressed some interest. Proponents of a fly-off believe that it would decrease the possibility of choosing the "wrong" technology and increase the likelihood of retaining competition in the domestic launch vehicle industry. Critics note, however, that a fly-off strategy would require larger near-term budgets than currently planned.

Another concern surrounding RLV development has been the role of DOD payloads and whether or not they will be used during the early testing of a commercial RLV. NASA and its commercial partners will need a sufficient number of payloads to both prove the reliability of RLV technology and attract potential investors. DOD officials, however, do not wish to repeat their negative experiences with the Space Shuttle and are, therefore, hesitant to contribute DOD payloads to the RLV until it is proven. Unless NASA and its industry partners can entice other payloads to fly aboard an RLV, DOD's reluctance could potentially drive up the price of launching payloads on the RLV.

ISSUE 1e: *SSTO?*

SSTO development entails significant technical risks. NASA has proposed a phased SSTO technology maturation program that periodically pauses along the way to determine the prudence of continuing. In the event the pursuit of SSTO is terminated at any point, NASA suggests that other RLV concepts (e.g., two-stages-to-orbit or TSTO systems) can then be considered and that new RLV efforts could possibly draw on past SSTO technology development efforts.

Some analysts and policymakers have taken issue with this approach, arguing that it grants too much upfront attention to the SSTO concept. They note that pursuing another RLV concept such as TSTO after investing significant resources in SSTO risks a suboptimal result that does not achieve the desired level of cost reductions.

Also, there has been some concern that NASA has not adequately defined its criteria for judging the success of the X-33 program. NASA, in conjunction with the Office of Science and Technology Policy and the Office of Management and Budget, has established criteria to support decisions in 1996 and 2000. Nevertheless, some analysts and policymakers are concerned that these criteria are insufficient and suggest that NASA provide Congress with a set of specific intermediate criteria for evaluating the success of the X-33 program on an annual basis. Such a requirement may, however, slow the development process.

Finally, although NASA has claimed that it must pursue block upgrades to the Space Shuttle if the government or industry decides in 2000 to forgo development of a commercial RLV follow-on to the X-33, there are other alternatives. For example, NASA could decide to extend SSTO development efforts, initiate TSTO development efforts, support development of a commercial TSTO, pursue block upgrades to the Space Shuttle, commission a new space transportation study, or reconsider alternative options already examined in past studies. Which alternative NASA chooses in coordination with its industry partners will depend on the progress made during the X-33 program, as well as the commercial prospects of an RLV.

ISSUE 1f: *Space Shuttle—beyond 2000*

As noted, NASA may decide in 2000 to pursue block upgrades to the Space Shuttle in order to ensure safe operations until 2020. Discussions with both NASA and industry officials reveal, however, that little planning for this possibility and little investigation into whether or not the industrial base will be able to support these upgrades are currently being done by NASA.

In its implementation plan, NASA has proposed replacing the existing solid rocket boosters (SRBs) with Liquid Fly Back Boosters (LFBBs) between 2007 and 2010. NASA claims that LFBBs would increase Shuttle safety, payload performance, and launch probability, and would also reduce annual Shuttle operating costs compared with SRBs. The implementation plan does not, however, outline any contingencies to address the significant impact that replacing SRBs with LFBBs might have on the solid rocket motor industry and the nation's continued ability to produce long-range ballistic missiles.

Finally, there remains the prospect of another Space Shuttle accident that results in the loss of an

Orbiter. Such a loss would have major repercussions for both the Space Shuttle and X-33 programs.

Fundamental Objective #2: U.S. Use of Foreign Launch Systems and Components

The NSTP encourages federal agencies to take advantage of foreign technologies in U.S. space transportation systems. It also limits the flight of U.S. government payloads to U.S. space transportation systems, in effect removing U.S. government payloads from the available international marketplace for launch services. In this, it follows past policy. In addition, the policy allows the launch of government payloads on foreign launch vehicles if they are made available on a no-exchange-of-funds basis and if they support cooperative scientific programs.

ISSUE 2a: *The use of foreign launch technology*

The use of foreign technologies in U.S. space transportation systems may improve the efficiencies of U.S. launch systems, assist U.S. access to space, and improve U.S. competitiveness in the international space transportation market. With the important exception of the Space Shuttle and its main engines, the United States has done relatively little launch system R&D since the 1960s. The use of foreign technologies in U.S. space transportation systems could reduce the amount of R&D now required of U.S. firms in efforts to improve the performance and reduce the costs of U.S. systems. Russian launch vehicles and related systems (particularly propulsion) have significant potential for commercial use. Russian hardware and space transportation skills can fill important gaps in U.S. capabilities. The United States might benefit from European space transportation technologies as well.

On the other hand, U.S. national security interests demand that the United States maintain a viable national launch capability and technology base. The use of foreign technologies might reduce the incentives for maintaining the domestic R&D that underlies that technology base.

The simple purchase of either vehicles or launch services appears to be less attractive than joint ventures, co-production of vehicles and/or systems, and analogous business arrangements, as ways of harmonizing these differing interests. For example, Aerojet and Pratt and Whitney, both U.S. manufacturers of liquid-fueled engines, are exploring ways in which to capitalize on the use of Russian liquid-fueled engines in U.S. vehicles.[6] U.S. launch operations experts have expressed interest in Russian and European methods to reduce operations costs. In its implementation plan, DOD has expressed openness to the use of foreign technologies in U.S. launch vehicles, but only under conditions that would protect the supply of critical components should foreign sources become unavailable. Each proposed technology insertion would be judged on a case-by-case basis.

Methods to protect component supply, such as stockpiling critical components or duplicating production lines in the United States would likely result in higher costs to the government, but might ensure that the United States will be able to fulfill its space-related national security needs in times of crisis. Officials of Arianespace have offered to sell the Ariane 5, or license rights to build it, to service U.S. HLV needs. Such an arrangement could substantially reduce the costs of building and operating a U.S. HLV.[7] However, building a vehicle under license might inhibit the development of new U.S. technology that could be used to improve the U.S. MLV fleet.

Experts disagree over the extent to which cooperation with the Russian government and industry on space projects would affect U.S. competitive-

---

[6] Michael A. Dornheim, "Aerojet Imports Trud NK-33 Rocket Engine," *Aviation Week and Space Technology*, Oct. 25, 1993, p. 29; and Jeffrey M. Lenorovitz, "Pratt Signs Accord with NPO Energomash," *Aviation Week and Space Technology*, Nov. 2, 1992, pp. 25-26.

[7] Ben Ionatta and Cheri Privor, "Arianespace's EELV Proposal Finds Little Favor," *Space News*, Apr. 10, 1995, p. 3.

ness and the retention of U.S. jobs. Some industry officials, for example, express concern that the United States could lose employment in the launch services industry if Russian technology were used extensively in U.S. launch systems. Others have argued that skillful incorporation of Russian technologies in U.S. systems could save taxpayer dollars in publicly funded programs like the International Space Station and boost U.S. international competitiveness in commercial programs. Greater competitiveness might generate new jobs in space transportation and space-related fields, partially or fully offsetting job losses due to the use of foreign technology.

Second-, third-, and fourth-tier launch system equipment suppliers appear to be most vulnerable to the extensive use of Russian technology in U.S. launch systems, especially those that now supply subsystems and parts for U.S. propulsion systems. Loss of critical skills in the lower tiers of the space transportation industrial base may, in some instances, adversely affect the nation's ability to maintain assured domestic access to space and reconstitute production of long-range ballistic missiles. Nevertheless, reducing the cost of access to space may well lead to more aerospace jobs as a whole.

Some observers worry that given the precarious state of the Russian economy and government, Russian equipment suppliers may not be able to sustain their ability to produce space goods and services.[8] Russian firms, recently privatized and undergoing massive restructuring, have experienced difficulties in moving to a demand-driven, market-oriented economy. Concern over the future ability of Russian firms to perform could be eased, in part, if Russian firms successfully demonstrate that they can produce goods and services on time and within the terms of cooperative contracts with the U.S. government and industry. The existing cooperative activities between NASA and the Russian Space Agency, especially with regard to construction and operation of the International Space Station, will provide considerable insight into the long-term viability of the Russian space transportation industry.[9]

The NSTP also allows the use of foreign launch systems to carry U.S. instruments and spacecraft on a no-exchange-of-funds basis when supporting cooperative programs with other countries. Examples of such cooperative use of non-U.S. launchers include the shipment of U.S. equipment to the Russian Mir space station aboard a Russian Spektr spacecraft launched on a Russian Proton launch vehicle in May 1995, and the earlier launch of the TOPEX/Poseidon spacecraft on an Ariane 4 in 1992. Such use can sharply reduce U.S. costs for science programs and may facilitate some projects that might otherwise not be flown, but could deprive U.S. launch providers of a few launch opportunities. The launching country gains by receiving access to data generated by the U.S.-built, or jointly built, instrument or spacecraft.

ISSUE 2b: *International trade in launch services*

In keeping with broader U.S. international trade principles, the NSTP seeks to achieve free and fair trade in launch services. However, as a result of the close connections between defense and launch system technologies, and the desire to achieve or retain autonomy in launch services, all spacefaring nations subsidize their launch services industry to some extent. Because the economic structure of each country is different, it is difficult to determine the true extent of the subsidy each extends to its launch industry. In addition, each spacefaring country generally reserves government payloads for its own launch systems. For example, in keeping with past U.S. policies, the NSTP requires that U.S. government payloads fly

[8] Judyth L. Twigg, "The Russian Space Program: What Lies Ahead?" *Space Policy* 10(1):19-31, 1994.

[9] U.S. Congress, Office of Technology Assessment, *U.S.-Russian Cooperation in Space*, OTA-ISS-618 (Washington, DC: U.S. Government Printing Office, April 1995).

on U.S. space transportation systems, except for well-defined cooperative programs.

Trade agreements with China and Russia, which are intended to manage the international market for launch services and reduce the impact of low Chinese and Russian prices on U.S. launch service companies, may also reduce international competition and raise the overall price of launch services. The United States first faced competition from non-U.S. launch service entities after ESA developed the Ariane launch system in the late 1970s. Specifically designed to carry payloads to geosynchronous Earth orbit (GEO), and marketed by the European corporation Arianespace, the Ariane system was designed and built on the premise that it would capture a significant share of the available world market in commercial payload launch services. Since the loss of the Space Shuttle *Challenger* in January 1986, Arianespace has garnered a dominant share of the international commercial payload market.[10]

During the late 1980s, China and Russia (then the Soviet Union) began to offer launch services on the international market, increasing the competitive pressure on the U.S. commercial launch services industry. Faced with growing competition in launch services, increasing concern that launch systems built in non-market economies would unfairly compete with U.S. launch systems, and pressure from U.S. satellite manufacturers to allow the launch of U.S.-built satellites on Chinese and Russian launch systems, the United States sought and obtained launch service agreements with China and Russia.

In addition to setting limits on the total number of Chinese and Russian launches within a specified period, the agreements attempt to establish rules by which the market will operate. The United States is able to exert influence over trade in launch services because it sells more satellites on the international market than any other country. Russia and China have signed trade agreements because the United States could severely restrict the international sale of U.S.-manufactured satellites launched on other countries' vehicles. The Office of the U.S. Trade Representative (USTR) is the U.S. agent in negotiating these agreements.

U.S. satellite manufacturers have begun to pressure the USTR to relax or do away with the existing restrictions on the number of Russian commercial launches allowed between now and the end of the century. They have been joined by U.S. partners of Russian launch companies, which would profit from relaxed restrictions. Existing and planned partnerships between U.S. and Russian companies are likely to complicate U.S. considerations of these agreements, making it much more difficult to assess overall benefits and drawbacks of changes in the agreements.[11] Relaxation of the U.S.-Russia agreement would make the launch services market more competitive. It might also undercut the ability of U.S. launch service providers to compete and indirectly raise the costs of space transportation services to the federal government.

Arianespace, which now commands the largest share of the commercial launch services market, may be more affected by a relaxation of the U.S.-Russia launch services agreement than U.S. firms. Although a relaxation of the agreement would increase the competitive pressures on U.S. launch companies not now associated with Russian companies (such as McDonnell Douglas, which markets the Delta MLV, and Orbital Sciences, which markets the Pegasus and Taurus SLVs), those companies launch payloads for the U.S. government and therefore would retain a strong core market for launch services. Lockheed Martin, which markets the Atlas ELV, also markets the Russian Proton

---

[10] Prior to the loss of *Challenger*, NASA actively marketed commercial launch services on the government-owned and -operated Space Shuttle. In August 1986, President Reagan issued a policy directive limiting the use of the Shuttle to payloads that required the unique capabilities of the Shuttle.

[11] Craig Covault, "Russian Proton Challenges Ariane," *Aviation Week and Space Technology*, Apr. 24, 1995, pp. 40-43.

through LKE International. Lockheed Martin intends to use the two vehicles to back each other up, should one be temporarily removed from service to correct a system failure.

ISSUE 2c: *Technology transfer and foreign policy objectives*

Cooperative ventures risk transferring domestic technologies that could be used to strengthen a competitor's position in the international aerospace market and to assist belligerent countries in developing the means of delivering weapons of mass destruction (nuclear, chemical, and biological weapons). Experts disagree over how effective means to prevent such transfer can really be, but present policy clearly moves toward loosening trade restrictions. For example, many items having to do with satellites and satellite technology have been moved from the U.S. Munitions List onto the Commerce Control List, effectively making it easier to trade in those items. Further loosening of restrictions could result in improved U.S. trade in space transportation technologies. On the other hand, the United States must also remain sensitive to the potential proliferation of missile-related technologies.[12]

U.S. cooperative agreements with other countries must conform with related U.S. obligations and treaties, such as technology transfer policies and the Missile Technology Control Regime (MTCR), which was developed in 1987 to limit proliferation of long-range delivery systems capable of delivering weapons of mass destruction.

Admittance to the U.S. satellite market has become a tool in encouraging adherence by China, Russia, and Ukraine to the MTCR. Russia and Ukraine have agreed to join the MTCR. The Clinton Administration believes that helping the Russian civilian space program stay as healthy as possible and capable of retaining its experts will reduce global proliferation of missile technology. China has declined to join the MTCR, but has agreed to abide by its restrictions. However, the United States has raised several issues of noncompliance with Chinese officials. On October 4, 1994, the United States and China agreed to "work together to promote missile nonproliferation through a step-by-step approach to resolve differences over missile exports."[13] The United States could levy sanctions against the Chinese launch company, including prohibition of satellite launches, if the United States found that the entity was selling missile-related technology to a country that did not previously possess such technology.

Fundamental Objective #3: The Use of Excess Ballistic Missiles

The NSTP reserves use of excess ballistic missiles for government payloads only, and only when their use results in cost savings to the government over the use of commercial launch services. Excess ballistic missiles can be used by the government for engineering tests and suborbital flights, but orbital flights that might compete with private launch services must satisfy tough conditions before they are allowed.

Some 650 long-range ballistic missiles will be made available by U.S. adherence to the first Strategic Arms Reduction Talks treaty alone. These missiles, and others to be retired under other treaties, could be used to launch government and commercial satellites into orbit. Even if the missiles themselves are not used, parts of the missiles and the tooling for building those parts could be useful to industry.

---

[12] U.S. Congress, Office of Technology Assessment, *Export Controls and Nonproliferation Policy*, OTA-ISS-596 (Washington, DC: U.S. Government Printing Office, May 1994).

[13] U.S. Department of State, Office of the Spokesman, "Joint United States-People's Republic of China Statement on Missile Proliferation," Fact Sheet, Oct. 4, 1994.

ISSUE 3a: *Unfair competition or market creation?*

The Clinton Administration's policy continues the Bush Administration's policy of tightly restricting the use of excess long-range ballistic missiles. Some analysts argue that making these missiles more widely available for use as space launch systems would not only save much needed government resources, but could also demonstrate the viability of new markets for SLVs. Others argue, however, that although such a scheme might save the taxpayers money in the short term, it might also drive commercial SLV vendors from the market, leaving the U.S. industry with no SLV producers in the long term.

There is a lack of data on how much it would cost to convert surplus ballistic missiles for new payloads, how useful these missiles might be for more delicate payloads, and how SLV providers might maintain their ability to develop new systems should converted ballistic missiles be priced below current SLVs. Those questions must be answered before the debate on how to use excess ballistic missiles can be resolved.

ISSUE 3b: *Russian excess ballistic missiles*

In contrast to American policy regarding surplus missiles assets, former Soviet Union firms are promoting a number of converted intercontinental ballistic missiles (ICBMs) and submarine-launched ballistic missiles for an assortment of commercial duties. Two already on the market are the Start-1 and the Rokot, derived from the SS-25 and SS-19 ICBMs, respectively.

Russia's use of its excess ballistic missiles as SLVs has not yet proven to be a viable commercial strategy. If Russia is successful in marketing its surplus ballistic missiles, however, U.S. SLV launch service providers will face international competition from Russian excess ballistic missiles, while the U.S. government will receive none of the benefits of selling its stockpiles.

Fundamental Objective #4: The Private Sector Role in Space Transportation Decisionmaking

The private sector is expected to play a crucial role in accomplishing many of the space transportation goals set forth in the NSTP and the supporting implementation plans. It is, for example, designated to be a source of: 1) significant funding in a fiscally constrained budget environment; 2) expertise to manage space launch activities more efficiently; and 3) innovative ideas and products in the design and development of future space transportation systems. Placing greater reliance on the private sector is in keeping with general trends that emphasize reducing government's responsibilities in areas in which the private sector might reasonably be expected to provide the desired goods and services.

But the private sector is not a monolithic entity with a single coherent view of space transportation needs or the goals outlined in the NSTP. While the principal prime contractors for space transportation are in general agreement on many aspects of the Clinton Administration's space transportation policy, they have different views about the implications of particular elements of policy. Additionally, some subtier firms are skeptical about the potential for the government to achieve the goals of the NSTP.

The willingness and, indeed, the ability of private sector firms to fulfill the roles suggested in the national space transportation planning documents depend in many instances on factors that possess a great deal of uncertainty and are difficult to estimate accurately (e.g., the size and character of the future commercial space transportation market) and that are highly dependent on actions by the government (e.g., the nature of any government-industry partnership). These facts raise several issues that Congress might wish to consider in evaluating the role that the private sector plays in implementing the Administration's policy.

ISSUE 4a: *Will the estimated market support policy goals?*

In the absence of a major increase in the government's space transportation budget, private sector investment is viewed as essential for the development and production of an RLV follow-on to the X-33 possessing the characteristics desired by NASA. But any private sector investment depends on the potential for sufficient return on that investment to make it attractive. The assessment of the size, character, availability, and relationship (potential overlap) of future space transportation markets is therefore critical to industry's attitude toward new launch vehicle development programs and government cost-sharing arrangements.

The current industry assessment of the space transportation market appears to be that the potential market for commercial payloads in the MLV class is by itself insufficient to entice enough private sector investment to build a future RLV capable of meeting NASA's needs.

Current analysis indicates that the government is likely to continue to be the largest single market for U.S. space transportation for at least the next 10 to 15 years. Expansion of the commercial market in areas such as communications and earth observation is probable, but the size and rapidity of such expansion is uncertain. A number of potential new markets, such as space manufacturing and tourism, are on the horizon, but the size and speed of development of these markets are uncertain. This uncertainty about future markets inhibits private sector investment.

Industry analysis indicates that the potential commercial market for small payloads may be sufficient to attract enough private sector investment to develop vehicles to meet both commercial and government needs for small payloads. The willingness of firms to invest in the X-34 program supports this conclusion.

U.S. industry sees little commercial need for heavy lift and views this as principally a government market. The private sector is unlikely to put much of its own funds in this area without strong government support. Some observers note, however, that Arianespace plans to replace its medium-lift Ariane 4 with the heavy-lift Ariane 5.

Such assessments imply that if the government desires an RLV replacement for the Space Shuttle, it will have to provide a significant amount of the funding—either through a direct development and procurement process or through some form of guaranteed business.

ISSUE 4b: *The nature of the government-industry space transportation partnership*

The NSTP and implementation plans stress the need for closer government-industry cooperation—what NASA terms a partnership. Government planners believe that a more important role for industry in decisionmaking is essential if industry is going to be asked to help finance much of the production of a future medium-to-heavy-lift RLV. There are, however, a number of questions about the nature of any new government-industry relationship and the possible implications of closer ties between the government and any particular firm.

There appear to be a number of advantages to closer cooperation between government and industry. One is a potentially more efficient and less costly management structure. Another benefit is more effective use of the nation's public and private sector space transportation industry's technical expertise and facilities. But closer cooperation raises serious questions about who decides what research topics to pursue, which efforts will be funded, who will own the technical data rights resulting from this partnership, and how these rights might be transferred if such transfer appears to benefit the government. These questions and many others will have to be addressed if a partnership is to be successful.

Each government organization appears to have different expectations for the government-industry relationship. The designated advocates of increased commercial participation are the DOT and DOC, but with little money and small staffs, these two Departments are likely to play a limited role. NASA needs private sector investment to build a new RLV. It is, therefore, interested in policies that will provide support for industry, as well as incentives for industry to invest. It has streamlined its program management, changed funding rules, and made its research staff available to industry.

DOD, in contrast to NASA, does not have the same perceived need for a new space launch vehicle to perform its missions. Its current capabilities are more costly and less flexible than desired, but they perform well enough to meet the Department's fundamental mission requirements. DOD therefore appears less concerned about developing a close partnership with industry than is NASA.

ISSUE 4c: *Risk management—striking the proper balance*

Uncertainties about the future space transportation markets increase the need for private sector firms to protect any investment against losses. With estimates on the cost of development and production of a future medium-to-heavy-lift RLV ranging from \$6 billion to \$20 billion, many in industry are supporting the concept of anchor tenancy (e.g., committing the federal government to purchase an agreed upon amount of launch services from commercial firms) as a means of encouraging industry to invest in RLV development and production. By providing a guaranteed market for a specific period, anchor tenancy would reduce investment risk for the private sector during the formation of a more robust commercial market. A recent example of a commercial anchor tenancy is McDonnell Douglas' agreement to develop a Delta III ELV in exchange for a commitment by Hughes Telecommunications and Space Co. to purchase 10 future launches.

There are a number of issues that must be addressed. One is that a program based on anchor tenancy might be considered a "lease-purchase" arrangement. This could make the arrangement problematic because current government accounting rules require that such an arrangement be recorded in the budget as if the government purchases the assets outright. The discounted value of the expected costs of space launch services would be recorded as budget authority when the contract was signed. Outlays would be recorded (scored) in proportion to the construction activity on the launchers, as if the government were building the system.

Other observers argue that there is a need for new thinking in anchor tenancy, particularly when the government is slated to be less involved in the development of goods and services that might come from the private sector. They argue that anchor tenancy might be successfully used if the situation is one in which there is little technological risk, the contractor is taking the risk of performance, the contractor is financing the project, and the contractor has design control. Competitive bidding to help establish the market assessment of risk is also important.

Several industry representatives have argued that the basis for an anchor tenancy arrangement needs to be established by April 1996, when industry must begin to commit significant funds toward the development of the X-33 technology demonstrator. Without this commitment, industry may still participate in the X-33 program, but will probably reduce its share of the investment.

Industry also argues that termination liability (e.g., requiring the government to compensate industry should the government cancel a launch contract for its own convenience) is essential for reducing the risk to the private sector of entering into a long-term launch service agreement with the government. Skeptics have argued that such arrangements amount to providing a "risk-free" environment for U.S. business. Still, termination liability usually does not provide for loss of future revenue, only for money already spent. Thus firms continue to risk the loss of future returns on the money invested and bear the opportunity cost of not having invested the money elsewhere, even if compensation for funds already spent is guaranteed.

ISSUE 4d: *Infrastructure*

Many analysts argue that significant launch cost savings might be realized through changes in launch operations and infrastructure. Some have suggested building new, more generic launch facilities. Many analysts indicate that important launch cost reductions are unlikely unless launch operations engineers and facility managers have a greater role in the design of future launch systems.[14] Efficient launch operations are key competitive advantages for both Arianespace and Russia. A future RLV may have a completely different launch infrastructure than that of ELVs.

Because of the importance of the impact of launch services and infrastructure to long-term government costs and commercial competitiveness, Congress may wish to pay particular attention to activities in these often overlooked areas. Questions of how future space transportation systems will operate and how such operations will save money in comparison with current operations might be key oversight issues.

ISSUE 4e: *Accommodating commercial needs*

Many in industry express concern over the extent to which development of new space transportation systems will be influenced by rigid government space launch and payload requirements rather than by accommodation of commercial space transportation competitiveness considerations. For example, although NASA has restructured its program management and made a number of procedural changes that can aid development, its program may still be best structured to produce an RLV that will serve the U.S. government's space transportation needs first—rather than producing a commercially viable vehicle that will also meet government needs.

Part of the problem is the NASA requirement to carry crews to and from the International Space Station. Another part of the problem is the inability to define what might be commercially viable. Some industry representatives have noted, for example, the need to design commercial vehicles to serve the GEO market. This might result in very different designs from those optimized for NASA's International Space Station mission. These issues will need to be resolved if the programs are to meet their objectives.

Additional Issues for Congress

Two important issues were not addressed by either the NSTP or its implementation plans, but warrant consideration by Congress. These are the preservation of long-range ballistic missile capabilities and the status of the lower tiers of the space transportation technology and industrial base.

ISSUE 5: *Preservation of long-range ballistic missile capabilities*

The U.S. Navy plans to procure the last long-range ballistic missile in the strategic nuclear arsenal in 2005. No plans currently exist to produce any additional missiles after that time. Without producing missiles, however, the United States' ability and capacity to design and produce long-range ballistic missiles will deteriorate unless significant efforts are made to preserve them.

Both the U.S. Air Force and the Navy have preservation programs underway, but they are limited to a small set of critical components. Solid rocket motor technology may be particularly threatened. At present, all U.S. long-range ballistic missiles use solid rocket motors. If both the EELV and RLV designs use only liquid-fueled engines, and if liquid-fueled boosters replace the Space Shuttle's solid rocket motors, the market for large solid rocket motors in the United States may all but disappear.

ISSUE 6: *The invisible lower industrial tiers*

Current policy focuses on the large prime contractors, but there is more to the U.S. space transportation industry than just those firms. Hundreds of smaller firms provide subsystems and components, to the extent that about 50 cents of every procurement dollar flows down to these lower tiers of the industry.

---

[14] See U.S. Congress, Office of Technology Assessment, *Reducing Launch Operations Costs: New Technologies and Practices*, OTA-TM-ISC-28 (Washington, DC: U.S. Government Printing Office, September 1988).

OTA found that many of the lower-tier firms are pessimistic about their chances of survival. They believe that the government is not committed to the actual completion of new launch vehicles, and that research and development money will not find its way past the prime contractors.

Congress may wish to consider what the chances are that some of these lower-tier firms might be forced out of business, and what effect that is likely to have on the United States' ability to compete in the international market. If all the companies that produce a particular component or material critical to the space transportation industry go out of business because of lack of funds from the upper-tier firms, it could be very difficult and expensive to regain the capability to produce that component again.

CRITICAL DECISION POINTS

Each of the space transportation policy implementation plans was accompanied by an idealized timeline. While each department and agency was careful to say that the timelines were not set in stone, they do provide policymakers with some sense of the important decisions that await them and some of the hidden problems they may face in a few short years. Table S-2 lists some of the more critical decision points and their potential implications. Changes in political leadership, new space program goals, stretched out or terminated programs, unforeseen technical difficulties, and launch failures are just a few events that could dramatically alter the timing of these important decisions.

TABLE S-2: Critical Decision Points and Their Possible Implications

Date	Decision or event	Possible implications
1995	Corporate investment strategies for X-33 and X-34 development programs must be formulated.	Corporate evaluation that government programs are unlikely to transpire as advertised could result in inadequate corporate investment, creating a self-fulfilling prophesy.
1996	Down-selection for Phase II of X-33 single-stage-to-orbit demonstration vehicle.	Industry participation in Phase II may require early government commitment to and legislative action on cost and risk sharing on the follow-on operational RLV. This is four years before NASA's specified 2000 decision on Space Shuttle upgrades (see below). Contract winner has an advantage for production of follow-on RLV, unless other companies invest in their own competitive vehicles or the contract winner fails to meet program performance and cost objectives.
1998	Down-selection to one EELV producer.	Contract winner will develop family of medium-to-heavy vehicles for DOD, perhaps consolidating U.S. ELV business to one firm. Med-Lite winner, however, may compete at lower payload range.
1999	X-34 RLV demonstration vehicle flight tests are completed.	A successful X-34 producer could potentially dominate the SLV market if significant per flight price reductions are achieved. U.S. government costs for SLVs drops. New markets may develop for LEO light satellites if X-34 producer drastically lowers per flight prices.
2000	Government decision required to pursue either a manned RLV variant or major block upgrades to the Space Shuttle.	A premature decision to develop a manned RLV could result in a less-than-revolutionary vehicle. Spending on major upgrades to the Shuttle could indefinitely postpone RLV development to the detriment of government launch expenditures and U.S. competitiveness in the commercial launch market.
2000+	Corporate decision to build a commercial RLV.	Size of government market and government commitment to RLV producer may lead producer to focus exclusively on government needs, at the expense of capturing and creating commercial markets. Alternatively, RLV producer may choose to construct two vehicles or a single vehicle with optional strap-on boosters to accommodate heavy government payloads and medium commercial payloads.
2001	Medium EELV becomes operational.	A maximum 10-percent cost savings from new MLV improves U.S. position in the commercial market, but not enough to hold off Russian and Chinese competition. To ensure success of its EELV program, DOD may avoid early participation on RLV flights, limiting the customer base for potential RLV investors.
2001/2	Trade agreements with Russia/ China expire.	Unless new agreements are negotiated, U.S. launch providers find themselves at a severe pricing disadvantage. Without an RLV or greater than expected savings from a medium EELV, launch providers may find themselves unable to compete in the commercial market.
2005	Heavy EELV becomes operational.	A maximum 40-percent savings on new HLV results in substantial cost savings to the government. Potential for the development of a multipayload version, like the Ariane 5, for limited, expensive commercial use.
2005	Limited RLV flights commence.	RLV begins direct competition for flights with the Space Shuttle, the EELV family, and other ELVs.
2005	Last of the current generation of long-range ballistic missiles is produced.	Lack of development or production programs may result in loss of ability to make new ballistic missiles without significant startup costs and delays. Move of industry to all liquid-fueled boosters on the Space Shuttle, EELV, and future RLVs would all but eliminate domestic production of large solid rocket motors.
2012	International Space Station is scheduled to cease operation just as a manned RLV replaces the Space Shuttle or block upgrades of the Shuttle commence.	One of the few planning goals identified for the RLV is its ability to deliver passengers and cargo to the space station orbit. Life extension of the International Space Station seems likely, especially if operations are passed to a commercial venture.

SOURCE: Office of Technology Assessment, 1995.

The National Space Transportation Policy: Issues for Congress

INTRODUCTION

For nearly four decades, the United States has relied on its national space transportation technology and industrial base to support a multitude of defense, intelligence, civil,[1] and commercial needs. This base has produced the rockets to explore the planets and land the first man on the moon; assure the nation's strategic nuclear deterrent; and supply policymakers with intelligence, geodesy, weather, navigation, and other important information. It has generated new technologies, supported new space-based businesses, and helped the U.S. balance of trade.

This vital industry, however, is currently being buffeted by events and trends that are forcing its major restructuring. The end of the Cold War has reduced the demand for defense and intelligence space launches and halted new development of long-range ballistic missiles. Federal budget deficit and debt concerns have put additional pressure on government departments and agencies to be more efficient and to reduce the number and cost of civil and national security space launches. And a debate has begun within the executive branch and Congress over the proper balance between public and private investment in space transportation.

In the commercial market, the aging domestic fleet of expendable launch vehicles (ELVs) is facing stiff global competition from the commercially focused European Ariane rocket, as well as from Russian and Chinese ELVs, which benefit from hidden

[1] Government other than national security.

U.S. DEPARTMENT OF DEFENSE

The Delta MLV is marketed by McDonnell Douglas and serves the lower end of the medium-sized payload market.

U.S. DEPARTMENT OF DEFENSE

The Altas MLV is marketed by Lockheed Martin and serves the upper end of the medium-sized payload market.

real costs and low-wage labor forces.[2] Meanwhile, entrepreneurs in the telecommunication, navigation, and remote sensing satellite industries project a moderate increase in demand for small and medium launch vehicles (SLVs and MLVs) to support their business ventures. Other potential space markets remain dormant, in part because of the high cost of access to space.

In response to these challenges, the Clinton Administration released a new National Space Transportation Policy (NSTP) on August 5, 1994.[3] The NSTP established the President's strategic vision for space transportation and directed the Department of Defense (DOD), the National Aeronautics and Space Administration (NASA), and the Departments of Transportation (DOT) and Commerce (DOC) to prepare detailed implementation plans supporting the new policy within 90 days.

---

[2] The mostly reusable Space Shuttle carried some commercial payloads before the Challenger accident in 1986. All other currently operational space transportation systems are ELVs. This report does not address suborbital launch systems or transportation systems designed primarily to move payload or passengers between or beyond Earth orbits.

[3] Presidential Decision Directive NSTC-4. Most, if not all, the text of this internal policy was released publicly in The White House, Office of Science and Technology Policy, "National Space Transportation Policy," Fact Sheet, Washington, DC, Aug. 5, 1994. See appendix for complete text of the fact sheet.

TABLE 1: Missions and Activities Dependent on Launch Services

Missions and activities	National security	Civil	Commercial
Intelligence collection and arms control monitoring	Current	N/A	Potential
Human space flight and life sciences	Potential	Current	Potential
Telecommunications and entertainment	Current	Current	Current
Weather observation	Current	Current	Current
Navigation	Current	N/A	Potential
Environmental monitoring	Potential	Current	Current
Geodesy	Current	Current	Current
Planetary exploration	N/A	Current	Potential
Photogrammetry	Current	Current	Current
Space sciences and astrophysics	Current	Current	Potential
Space manufacturing and materials science	Potential	Current	Potential
Hazardous waste disposal	Potential	Potential	Potential

SOURCE: Office of Technology Assessment, 1995.

This report examines the policy and supporting implementation plans and raises issues that Congress may wish to consider as it debates the future of U.S. space transportation. It is structured around the four fundamental objectives identified by the Administration when it announced its new policy.[4]

The U.S. Space Transportation Technology and Industrial Base

The U.S. space transportation technology and industrial base is the vast and complex collection of people, institutions, technological know-how, and facilities needed to conceptualize, design, develop, test, produce, operate, and maintain space transportation systems and their supporting infrastructure.[5] This base spans government, academia, and industry from the largest prime contractor to the smallest commodity supplier.

This is not only a prestigious industry involving many high-skilled jobs, but also one entwined with other high-technology industries that depend on access to space or technologies developed for space.[6] It is vital to many intelligence and defense missions. (See table 1.)

The U.S. space transportation technology and industrial base provides launch services to government and commercial customers, as well as long-range ballistic missiles to DOD. Historically, the U.S. government has been, and still is, the largest customer for U.S. launch services, buying launches for defense, intelligence, and civil pur-

[4] The White House, Office of Science and Technology Policy, "Statement on National Space Transportation Policy," Washington, DC, Aug. 5, 1994. See appendix for complete text of the statement.

[5] As a part of its current assessment, OTA will publish a detailed report on the long-term prospects for the space transportation technology and industrial base in 1996.

[6] Space transportation services accounted for less than 10 percent of the $6.5 billion commercial space market in 1994. See U.S. Department of Transportation and U.S. Department of Commerce, "National Space Transportation Policy Implementation Plan," submitted to the White House Nov. 7, 1994, but not yet validated, p. 4.

poses. In 1994, it procured about 85 percent of all domestic space launches to meet national security or scientific objectives. About half of these launches involved the most expensive U.S. space transportation systems: the Titan IV and the Space Shuttle.[7]

The large government role in the U.S. space transportation technology and industrial base has had a major effect on its character and culture. Like many defense-oriented industries, the space transportation industry has operated in a business environment significantly different from that of purely commercial businesses.[8] The regulations and controls that have been a part of doing business with the government have contributed to the slowness of the U.S. response to growing international competition in the commercial market. Moreover, the relatively small number of launches worldwide open to international commercial competition has been less attractive than the sizable U.S. government market.

Some analysts predict the deployment of a new generation of smaller satellites, forming low-Earth-orbit (LEO) constellations that will enable voice, video, data, and multimedia services around the globe. They believe these satellites will spark a telecommunications revolution and thus spur new SLV development to meet the growing demands of a competitive marketplace. Moreover, some analysts see a number of other new markets developing, particularly for remote sensing and global positioning/navigation services. And a few analysts foresee even greater market opportunities in the more distant future (e.g., space manufacturing, tourism, nuclear waste disposal, and life-science research), if space transportation were made safer, more reliable, and much less costly.[9]

TABLE 2: 1994 World Orbital Launch Vehicles

Country	Vehicle	lbs to LEO	lbs to GTO[a]	1994 flights
United States	Scout	500	N/A	1
	Pegasus	620[b]	N/A	3
	Taurus	2,100[b]	N/A	1
	Titan II	4,200[b]	N/A	1
	Atlas E	1,750[b]	N/A	2
	Delta II	11,110	4,010	3
	Atlas I	13,000	5,240	2
	Atlas IIA	16,050	6,970	2
	Atlas IIAS	19,050	8,450	1
	Titan IV	39,100	14,000	4
	Shuttle	53,500	13,000	7
Russia	Rokot	154[c]	N/A	1
	Start-1	1,000	N/A	0
	Kosmos	3,000	N/A	5
	Tsyklon	8,800	N/A	8
	Molniya	N/A	N/A	3
	Soyuz	15,400	N/A	15
	Zenit	30,300	N/A	4
	Proton	44,100	12,100	13
France	Ariane 42P	13,200	5,730	2
	Ariane 42L	16,300	7,050	1
	Ariane 42LP	18,300	8,160	3
	Ariane 44L	21,100	9,260	2
China	Long March 2D	N/A	2,750	1
	Long March 3A	15,800	5,500	3
	Long March 2E	19,400	7,430	1
Japan	H-II	22,040	8,816	2
India	ASLV	330[b]	N/A	1
	PSLV	6,610	990	1

[a] GTO = Geosynchronous transfer orbit, a temporary orbit used to reposition spacecraft into geosynchronous Earth orbit (GEO).

[b] Polar orbit.

[c] Demonstrated.

SOURCE: Bretton S. Alexander et al., *1994 Space Launch Activities* (Arlington, VA: ANSER, January 1995). Office of Technology Assessment, 1995.

[7] Bretton S. Alexander et al., *1994 Space Launch Activities* (Arlington, VA: ANSER, January 1995).

[8] For a discussion of how defense and commercial industries differ, see U.S. Congress, Office of Technology Assessment, *Assessing the Potential for Civil-Military Integration: Technologies, Processes, and Practices*, OTA-ISS-611 (Washington, DC: U.S. Government Printing Office, September 1994).

[9] See, e.g., T.F. Rogers, "Toward a New Public-Private Space Transportation Strategy," *The Journal of Practical Applications in Space* 5(1):1-41, 1993.

Today, six major centers offer space transportation services: Russia,[10] the United States, Western Europe, Japan, the People's Republic of China (hereafter China), and India. Table 2 lists the primary orbital launch systems in operation in 1994.

The rise of a competitive, global market for commercial launch services in the past decade has cost U.S. launch service providers their previous monopoly on launching commercial satellites.[11] In 1979, the European Space Agency (ESA) successfully launched its first Ariane 1 rocket and began to compete for commercial payloads. Meanwhile, the U.S. government began to promote the Space Shuttle for commercial and government payloads by charging below-market prices, and in the process weakened domestic ELV producers.

The U.S. share of the world commercial space transportation market, however, plummeted in 1986 after the explosion of the Space Shuttle *Challenger* and did not begin to recover until about 1990, when payloads designed to be carried aboard the Space Shuttle were reconfigured for launch on domestic ELVs and DOD investments in the MLV I and MLV II lowered Delta and Atlas prices to rates competitive with Ariane.[12]

Still, current U.S. MLV launch vehicles are based on decades-old designs and are generally less competitive than foreign launchers in terms of price, launch schedule, launch operations, regulatory environment, and launch facilities.

ESA developed the Ariane 4 MLV and the forthcoming Ariane 5 heavy launch vehicle (HLV) to serve not only the launch needs of ESA and its member countries, but also the commercial market at large. Commercial competitiveness played an important part not only in the design and sizing of the Ariane family of vehicles, but in launch infrastructure and operations that are optimized to meet commercial payload needs and schedules.

ARIANESPACE

Arianespace's Ariane 4 currently dominates the commercial market for launching medium-sized payloads.

More recently, China and Russia have entered the commercial launch market, confronting U.S. launch providers with non-market economy competitors who are able to undercut U.S. launch bids significantly even under the terms of existing launch service trade agreements.[13]

[10] Russia has launch sites in Russia and Kazakhstan. Ukraine is on the verge of entering the world space transportation services market.

[11] Some analysts argue that there is not a true global commercial market today, because of various government subsidies by the U.S. and foreign governments.

[12] All commercial payloads that had been booked for the Space Shuttle from 1987 to 1989 were flown on the Ariane, giving Ariane 100 percent of the commercial market for that period.

[13] These trade agreements do, however, set quotas on the number of Russian and Chinese vehicles winning commercial competitions. These agreements are discussed in detail in the section covering fundamental objective #2.

The United States remains the world commercial market leader in SLVs. These small rockets, however, account for relatively few of the total dollars spent on space transportation.

At the same time, the end of the Cold War, the desperate state of the Russian economy, and the Russian interest in developing a market economy have opened numerous opportunities for cooperation in space transportation technologies. Several American companies are negotiating with their Russian counterparts to secure Russian space launch technology, expertise, and hardware (e.g., engines). The recently merged Lockheed Martin Corp. now finds itself in the interesting position of being a partner in a U.S.-Russian joint venture (LKE International) to sell Russian launch services that compete directly with its own U.S.-made vehicles.[14]

The degree to which current commercial trends create problems for the United States space transportation effort is a matter of dispute. Some analysts argue that U.S. government requirements are sufficient to maintain a viable space transportation technology and industrial base and suggest that the government should not be overly concerned by the global competitive position of U.S. launch providers.[15]

Other analysts, however, believe that the development of new, low-cost space transportation systems must be encouraged to enhance the competitiveness of the U.S. space transportation industry and other domestic industries that require access to space. In their view, the government will also benefit from lower vehicle costs brought about by new technologies and larger sales volumes. With sufficiently low prices, the space transportation industry will be able to capture greater commercial market share. And domestic satellite vendors will not find themselves dependent on the possible vagaries of foreign launch providers. These analysts would like to see commercial market interests actively considered when the government embarks on development of a next-generation space transportation system:

> The future space transportation system selected must be responsive to commercial user requirements in addition to those of government users. While low operating cost is fundamental, other parameters, such as launch dependability, higher reliability, very short booking time, and user friendliness, are of equal importance. Another commercial requirement that will eventually emerge is the ability to accommodate the general public (in space flight) without rigorous astronaut-type training. These varied requirements and systems capabilities must be introduced in the current technology development plans. Unless the next space transportation system satisfies these needs, that system will not be widely used commercially.[16]

In general, however, industry officials are skeptical about the prospects for the development of new, low-cost space transportation systems. In the past decade, several new and often revolutionary proposals have been put forth, only to be dropped. These included the Advanced Launch System, National Launch System, the Air Force Space Lifter, the Space Shuttle C, the Space Shuttle II, and the National Aerospace Plane (NASP). This skepticism seems particularly profound within the lower tiers of the space transportation industry.[17]

---

[14] LKE International President Charles Lloyd says that the Russian Proton and American Atlas rockets could serve as backups for each other. See Patrick Seitz, "Lockheed Martin Corp. Officials Ready Ax," *Space News*, Mar. 20, 1995, p. 6.

[15] See, e.g., U.S. Department of Defense, Office of the Deputy Assistant Secretary of Defense, "Industrial Assessment for Space Launch Vehicles," Washington, DC, January 1995, p. ES-1.

[16] The Commercial Space Transportation Study Alliance, "The Commercial Space Transportation Study: Executive Summary," April 1994, p. 18.

[17] OTA conducted a workshop on the lower tiers of the space transportation technology and industrial base on March 2, 1995. Findings of this workshop will be presented in a short background paper to be released in June 1995.

U.S. DEPARTMENT OF DEFENSE

The multi-warhead Peacekeeper ICBM could be removed from the U.S. strategic arsenal under future arms control agreements and converted for space launch.

Finally, long-range ballistic missiles constitute an important but often overlooked segment of the space transportation technology and industrial base. Originally, most space launch vehicles were derived from ballistic missiles or ballistic missile technology. Today, these two segments of the industrial base have diverged significantly, but important overlaps in both technology and business arrangements remain. Therefore, reductions in the numbers of long-range ballistic missiles under various arms control agreements and the planned cessation of production in 2005 may have an important impact on the space transportation industrial base. Similarly, future choices in launch vehicle technologies may have important consequences on future capabilities to reconstitute long-range ballistic missile design and production.

The National Space Transportation Policy and Its Implementation Plans

The origins of the new space transportation policy and its implementation plans can be traced to the congressional request in the FY 1993 VA-HUD-Independent Agencies Appropriation Act for NASA to study government space launch needs and NASA plans for space transportation and the International Space Station. The "Access to Space Study," published by NASA in January 1994, responded to the congressional request as well as to NASA's own internal planning needs.[18]

Subsequently, Congress requested a similar assessment by DOD in its 1994 Defense Authorization Act (see appendix A). DOD's 1994 "Space Launch Modernization Study," more commonly known as the Moorman Report, responded to this congressional mandate.[19] This report built on previous studies, including the 1990 Augustine Report,[20] the 1992 Aldridge Study,[21] DOD's "Bottom-Up Review" in 1993, and NASA's "Access to Space Study".

The "Access to Space Study" and the Moorman Report provided the opening Department and Agency positions in interagency negotiations that led to the new Administration policy.

The NSTP released by the White House on August 5, 1994, was developed by the National Science and Technology Council (NSTC) and approved by President Clinton after an extensive

[18] NASA, "Access to Space Study: Summary Report," Washington, DC, January 1994.

[19] U.S. Department of Defense, "Space Launch Modernization Study: Executive Summary," Washington, DC, 1994.

[20] "Report of the Advisory Committee on the Future of the U.S. Space Program," Independent Report for NASA, December 1990.

[21] Vice President's Space Policy Advisory Board, "The Future of U.S. Space Launch Capability," Washington, DC, November 1992.

interagency review of current and future space transportation plans and budgets.

The policy provides overall guidance and direction for the executive branch. The specifics for implementing the policy were left to a second round of negotiations between the individual departments and agencies and the Office of Science and Technology Policy (OSTP). The policy required that DOD, NASA, and DOT/DOC (in combination) submit implementation plans within 90 days.[22] The DOD and NASA implementation plans were submitted to the White House in November 1994 and approved in November 1994 and January 1995, respectively. The DOT/DOC implementation plan was also submitted in November 1994, but has not received final approval.

MEETING THE FUNDAMENTAL OBJECTIVES OF THE POLICY

The Clinton Administration's four fundamental objectives for the NSTP were to establish new national policy regarding:

1. federal space transportation spending, consistent with current budget constraints and the opportunities presented by emerging technologies;
2. federal agencies' use of foreign launch systems and components;
3. federal agencies' use of excess ballistic missile assets for space launch, to prevent adverse impacts on the U.S. commercial space launch industry; and
4. an expanded private sector role in the federal space transportation R&D decision making process.[23]

In the following sections, the Office of Technology Assessment (OTA) examines each of these fundamental objectives and how the DOD, NASA, and combined DOT/DOC implementation plans propose to achieve them. The report then raises several issues of potential interest to Congress as it considers funding and oversight of the Administration's space policy, as well as changes in the laws governing space transportation.

Fundamental Objective #1: Space Transportation Funding and the Division of Responsibilities

The first objective of the NSTP is to set government spending priorities for current and future space transportation systems by assigning specific roles and functions to designated departments and agencies. The policy assigns to DOD the responsibility of overseeing improvements to the existing fleet of ELV systems and guiding development of new ELVs. NASA will continue to support the Space Shuttle and undertake research and development (R&D) that could lead to a new, reusable launch vehicle (RLV), replacing the Space Shuttle.

FUNDAMENTAL OBJECTIVE #1

Establishes new national policy for federal space transportation spending, consistent with current budget constraints and the opportunities presented by emerging technologies. Under the new policy, DOD will assume the lead responsibility for modernization of the current expendable launch vehicle fleet. NASA will assume lead responsibility for research and development of next generation reusable systems. NASA will focus their investments on technologies to support a decision no later than December 1996 on whether to proceed with a flight demonstration program. This program would, in turn, provide the basis for deciding by the end of the decade whether to proceed with a new launch system to replace the aging shuttle fleet.

This section examines the proposed division of responsibility between DOD and NASA and the development of new launch vehicle systems proposed in the DOD and NASA implementation plans. This examination suggests that, while the Clinton Administration's attempt to eliminate conflicts and redundancies may be a step in the

[22] The White House, Office of Science and Technology Policy, op. cit., footnote 3, sec. VIII(1).

[23] The White House, Office of Science and Technology Policy, op. cit., footnote 4.

right direction, several potential conflicts and redundancies remain. The examination further suggests that a number of additional concerns with the DOD and NASA implementation plans—especially the proposed development programs—also remain.

DOD

DOD's implementation plan initiates an ELV development program known as the Evolved Expendable Launch Vehicle (EELV) program. The EELV program would draw on existing launch technology to build a family of MLVs and HLVs that uses common components and subsystems. Table 3 shows the payload delivery capacities for both existing and proposed international space transportation systems.

By developing the EELV family of launch vehicles, DOD hopes to bring down the cost of putting its payloads into space, especially in the heavy payload class. Compared with current expenditures, DOD officials expect to achieve a 25- to 50-percent reduction in overall life-cycle costs for launching DOD payloads.[24] DOD anticipates that the EELV program will achieve these savings through the acquisition process, increased use of commercial products and practices, minimal use of unique government specifications, maximum use of common components and subsystems, and decrease in launch infrastructure costs.

Specifically, DOD believes that using common components and subsystems will, for the most part, eliminate the need to maintain an independent HLV production line and will also result in larger production volumes, possibly capturing economies of scale. The Department expects that the EELV program will benefit from current industry consolidation and reduce the need to maintain unique launch infrastructure and operations crews for multiple types of launch vehicles.

MCDONNELL DOUGLAS

Test flights of the DC-X RLV demonstrated the potential for rapid turnaround and simplified operations.

Why EELV?

DOD is—and projects that it will continue to be—a primary user of ELVs in all payload classes.[25] DOD officials believes that space transportation is too costly and note that DOD currently spends roughly $1.9 billion per year for space launch services. Of this money, $1.6 billion goes to support

[24] U.S. Department of Defense, personal communication, March 1995.

[25] U.S. Department of Defense, op. cit., footnote 15, p. II-11.

TABLE 3: Existing and Proposed International Space Transportation Systems

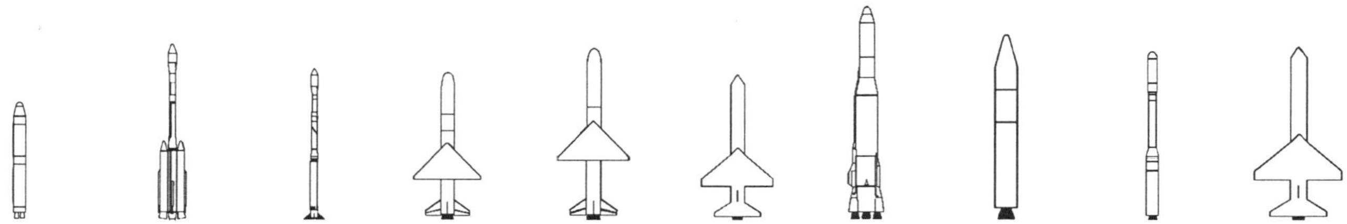

Vehicle	Rokot	ASLV	Scout	Pegasus	Pegasus XL	X-34A[c]	Atlas E	LLV1[e]	Taurus	X-34B[c]
Class	Small	Small	Small	Small	Small	Small	Small	Small	Small	Small
Country	Russia	India	U.S.	U.S.	U.S.	U.S.	U.S.	U.S.	U.S.	U.S.
lbs. to Leo	154[a]	330[b]	560[b]	731[b]	1.021[b]	1,200[d]	1,750[b]	2,000	2,100[b]	2,500[d]
lbs. to GEO	N/A	N/A	N/A	N/A	N/A	N/A	N/A	N/A	N/A	N/A

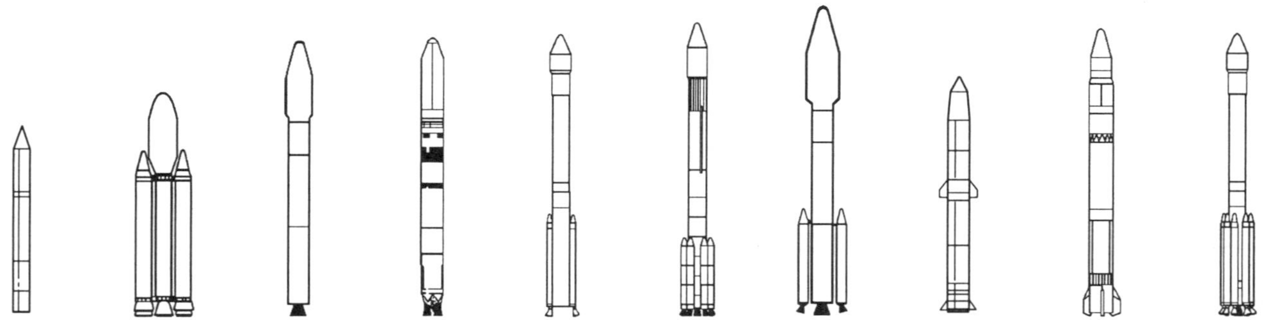

Vehicle	Kosmos	Conestoga[e]	LLV2[f]	Titan II	Med-Lite[f]	PSLV	LLV3[f]	Tsyklon	LM-2D	Delta II
Class	Small	Small	Small	Small	Small	Medium	Medium	Medium	Medium	Medium
Country	Russia	U.S.	U.S.	U.S.	U.S.	India	U.S.	Russia	China	U.S.
lbs. to Leo	3,000	3,300	4,000	4,200	4,790	6,610	8,000	8,800	N/A	11,110
lbs. to GEO	N/A	N/A	N/A	N/A	N/A	990	N/A	N/A	2,750	4,010

[a] Demonstrated.

[b] Polar Orbit.

[c] Initially, either the X-34A or X-34B will be developed, not both. The X-34 development team, led by Orbital Sciences and Rockwell International, will decide which version of the X-34 to build later in 1995.

[d] Orbit of 150 nautical miles (nmi) at an inclination of 28.5.

[e] First flight scheduled for 1995.

[f] Under development.

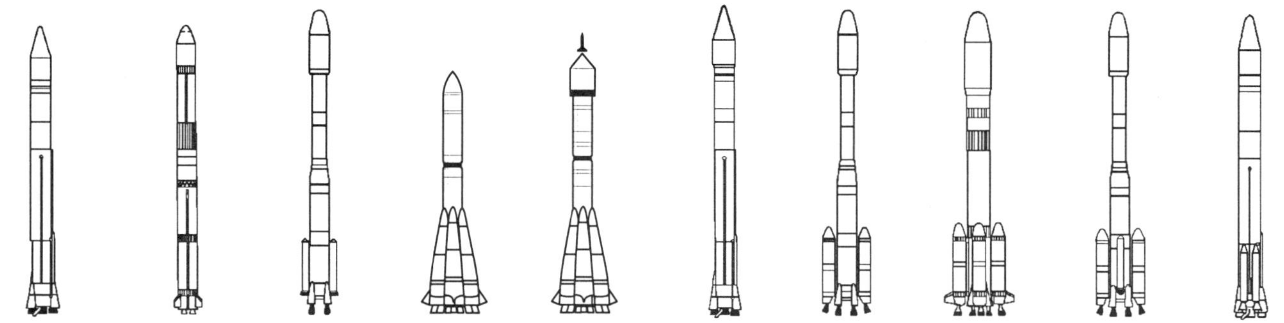

Vehicle	Atlas I	LM-3A	AR 42P	Molniya	Soyuz	Atlas IIA	AR 42L	LM-2E	AR 44LP	Atlas IIAS
Class	Medium	Medium	Medium	Medium	Medium	Medium	Medium	Medium	Medium	Medium
Country	U.S.	China	France	Russia	Russia	U.S.	France	China	France	France
lbs. to Leo	13,000	15,800	13,200	N/A	15,400	16,050	16,300	19,400	18,300	19,050
lbs. to GEO	5,240	5,550	5,730	N/A	N/A	6,970	7,050	7,430	8,160	8,450

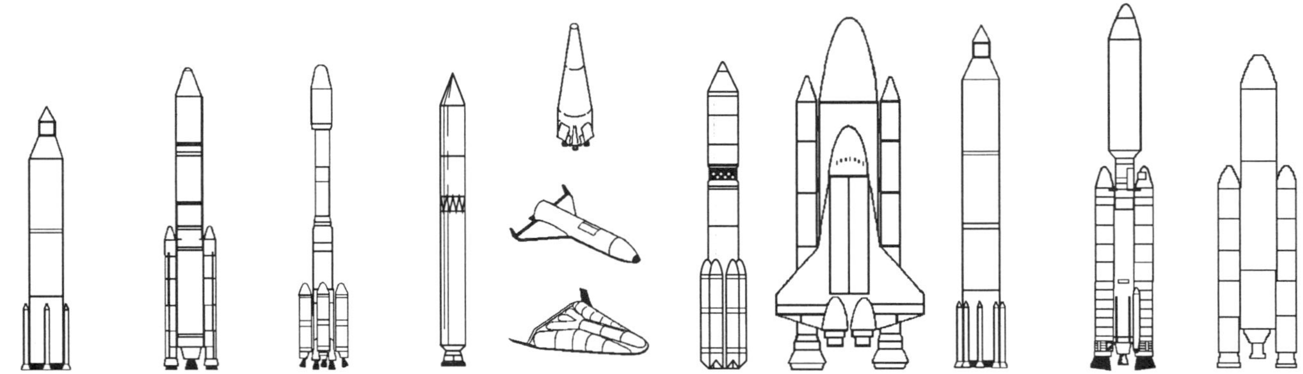

Vehicle	M-EELV[f]	H-II	AR 44L	Zenit	X-33[h]	Proton	Shuttle	H-EELV[f]	Titan IV	Ariane 5[e]
Class	Medium	Medium	Medium	Medium	Medium	Heavy	Heavy	Heavy	Heavy	Heavy
Country	U.S.	Japan	France	Russia	U.S.	Russia	U.S.	U.S.	U.S.	France
lbs. to Leo	No Spec[g]	22,040	21,100	30,300	40,000[i]	44,100	53,500	41,000[b]	39,100	44,000
lbs. to GEO	No Spec[g]	8,816	9,260	N/A	N/A	12,100	13,000	13,500	14,000	15,000

[g] DOD has not specified payload delivery requirements for the medium EELV. The EELV manufacturer could potentially develop more than one medium-sized vehicle to handle the wide range of DOD payloads in the medium payload class.

[h] Three development teams led by Lockheed Martin, McDonnell Douglas, and Rockwell International respectively, are conducting preliminary X-33 studies for NASA. At present, NASA plans to select one development team in December 1996 to build and test-fly its version of the X-33.

[i] This is equivalent to delivering 25,000 lbs. to the International Space Station orbit. According to NASA, the 25,000 lbs. to the International Space Station orbit is for planning purposes only and could be revised as development proceeds.

SOURCES: Bretton S. Alexander et al., *1994 Space Launch Activities* (Arlington, VA: ANSER, Jan. 1995). Office of Technology Assessment, 1995.

TABLE 4: Summary of Options Presented in DOD's 1994 Space Launch Modernization Study

	Option 1 Sustain existing launch systems	Option 2 Evolve expendable launch vehicles	Option 3 Develop new expendable launch system	Option 4 Develop new reusable launch system
Key features	Maintain the current fleet of launch systems for the foreseeable future Only upgrade to enable missions, improve reliability and safety, or address obsolescence Allow market-driven industry downsizing in order to reduce operations costs from current levels	Fly current launch vehicles already on contract Evolve a family of launch vehicles from current systems by consolidating medium- and heavy-lift booster families Lower operations costs by increasing production rates Cost: $1-2.5B	Correct deficiencies in current ELV fleet Significantly improve reliability, operability, and cost Develop a modular family comprised of a common core vehicle and/or common major subsystems Option A: replace existing ELVs Option B: replace existing ELVs and Space Shuttle	Substantially reduce flight costs while improving operability and responsiveness Tech. dev./demo. Costs: $0.6-0.9B Engr./dev. costs: $6-20+B Procure fleet of four vehicle between 2004-2009: $2.5-$10.5B Future operations costs: $0.5-1.5B per year
Projected costs per flight	Medium: $50-125M Heavy: $250-320M	Medium: $50-80M Heavy: $100-150M	Medium: $40-75M Heavy: $80-140M Personnel: $90-190M Cargo: $130-230M	None projected

SOURCE: U.S. Department of Defense, "Space Launch Modernization Study," 1994.

DOD's HLV program, the Titan IV.[26] The remaining $300 million is spent primarily on MLVs, with some of this remaining money going to SLVs. DOD has focused on cost reduction as its primary objective, especially in the heavy payload class.[27]

DOD's 1994 "Space Launch Modernization Study" offered a host of options for modernizing DOD's ELV fleet in a manner that would bring down the cost of access to space, especially for heavy payloads (see table 4). The EELV program is the implemented version of Option 2 presented in the study.

EELV Schedule and Funding

The EELV program schedule, which spans the period FY 1995-2005, is divided into three phases (see table 5). During Phase 3, DOD expects MLV development to precede HLV development. DOD projects that the MLV will become operational in 2001, while the HLV will become operational in 2005. DOD officials claim that this staggered introduction is likely to make the EELV procurement more attractive to potential producers since a new MLV will probably improve industry's com-

[26] The production cost of a Titan IV—although not firmly established—is on the order of $250 million per vehicle. Assuming an average launch rate of three Titan IVs per year, $750 million goes to the actual launch vehicles and the remaining $850 million goes to support payload integration, operational staff, and other supporting infrastructure.

[27] DOD is currently investigating ways to reduce its requirement for HLVs.

TABLE 5: Proposed EELV Schedule

Phase	Name	Duration	Expected number of contractors
1	Low-cost concept validation[a]	Aug. 1995-Nov. 1996	3-4[b]
2	Preliminary engineering, management, and development	Dec. 1996-Apr. 1998	2
3	Engineering, management, and development	May 1998-2005	1

[a] DOD has scheduled this 15-month period to allow industry to demonstrate, with about 90-percent confidence, that it can achieve EELV cost reductions targets within the proposed $2.017 billion budget. U.S. Department of Defense, personal communication, March 1995.

[b] DOD has allotted $120 million in contract money for this phase. DOD plans to award four contracts for this phase, with each team receiving $30 million for 15 months of work. DOD is exploring, however, the idea of awarding only three contracts for this phase, which would add $10 million to each contract. DOD will determine how many contracts to award for this phase based on whether the Department believes the $10 million add-on per contract will produce significant added value or not. U.S. Department of Defense, personal communication, March 1995.

SOURCE: U.S. Air Force, "EELV Briefing," *Space Day*, Colorado Springs, CO, Dec. 15, 1994.

TABLE 6: Proposed EELV Funding[a]

Fiscal year	Funding ($M)
1995[b]	40
1996	111
1997	72
1998	104
1999	173
2000	107
2001	110
2002	250
2003	460
2004	330
2005	260
TOTAL	**2,017**

[a] FY 1995 dollars.

[b] Includes $30 million appropriation for FY 1995 and redesignation of $10 million from FY 1994.

SOURCE: U.S. Air Force, "EELV Briefing," *Space Day*, Colorado Springs, CO, Dec. 15, 1994.

petitiveness in the commercial space launch marketplace. Table 6 provides the proposed EELV funding for FY 1995-2005.

Existing ELV Upgrades

When DOD began to formulate its EELV development program, it had several ELV upgrade programs underway. Since its decision to pursue the EELV, DOD has sought to limit redundancies by cutting back on ELV upgrade programs—retaining those programs that it believes are necessary to keep the existing ELV fleet flying safely or have significant short-term payoff (see box 1).

NASA

In its implementation plan, NASA outlines an RLV development strategy consisting of two X-vehicle programs—the X-33 Advanced Technology Demonstrator and the X-34 Small Reusable Booster. The X-33 program is designed to prove the feasibility of a medium-size, reusable, single-stage-to-orbit (SSTO) launch vehicle, while the X-34 program focuses on providing early test experience with a variety of technologies projected to go on commercial RLVs, as well as early experience with government-industry cooperative endeavors. The X-34 program is also expected to lead to a small, commercially operated, partially reusable launch vehicle. The impetus behind both X-vehicle programs is the belief that technology has advanced sufficiently to pursue development of reusable space transportation systems that could dramatically lower the cost of access to space.

NASA plans to spend just under $1.2 billion between FY 1995 and FY 1999 on its two X-vehicle programs. The X-33 program will receive

BOX 1: ELV-Related Upgrade Programs at DOD

Although DOD has scaled back its long-term efforts to upgrade expendable launch systems, several ELV-related upgrade programs with short-term benefits are still underway. Ongoing DOD programs include the Range Standardization and Automation (RSA) program and vehicle and infrastructure upgrade programs for the Titan II, Delta II, Atlas II, and Titan IV.

DOD views RSA as an essential upgrade to existing range operations infrastructure and expects the benefits to apply to all future launch systems—whether they be upgraded versions of existing ELVs or a family of EELVs. DOD's implementation plan notes that the RSA program—which includes significant computer and electronics upgrades—will overhaul operations at both the Eastern and Western ranges, and that DOD expects the program to be completed by FY 2004 at a cost of more than $1 billion.

Also, according to DOD's implementation plan, the Department will continue to operate several expendable launch systems and intends to complete ongoing development programs for these vehicles. The Delta launch vehicle flight safety and avionics upgrades are scheduled for completion in FY 1996. DOD expects that upgrades to the Atlas II propulsion system aimed at improving the launch vehicle's reliability will be completed in FY 1998.[1] And near-term Titan IV initiatives to improve reliability, enhance schedule dependability, and lower life-cycle costs are ongoing.

With regard to infrastructure, DOD's implementation plan states that the Department will maintain its launch capabilities for the Delta, Atlas, and Titan IV at Cape Canaveral. At Vandenberg Air Force Base, Titan II and Titan IV launch capability will be sustained, the construction of a new Atlas II Space Launch Complex will continue, and the Delta launch complex will still be jointly supported by DOD and NASA. The new Atlas II launch facility at Vandenberg is projected to be ready for NASA and national security missions in FY 1998.

[1] This refers to continuing development efforts of an RL-10C upper-stage engine. DOD does not currently have plans to incorporate a newly developed RL-10C engine in the Atlas launch vehicle.

SOURCE: Office of Technology Assessment, 1995.

$662 million of this funding, with an additional $339 million going to fund "Supporting Technology Demonstration" and other RLV-related technology development programs.[28] The X-34 program—which will span a shorter period of time than the X-33—will receive a fixed NASA contribution of $70 million. Both the X-33 and X-34 programs will likely draw from the operational and technological experiences of NASA's ongoing DC-XA program (see box 2).

Declining space transportation budgets have prompted NASA to pursue a strategy of close cooperation with industry in the development of reusable launch systems. NASA outlined this strategy in the cooperative agreement notices (CANs) it issued for each X-vehicle program. Each CAN delineates NASA's cooperative development policies and guidelines and includes the proposed funding contribution from NASA. NASA expects industry to contribute funding to each program because it believes that the X-33 and X-34 programs will help industry build RLVs capable of competing in the launch services mar-

[28] "Supporting Technology Demonstration" and other RLV-related technology development programs are sometimes collectively referred to as the "core technology development programs" for the RLV.

BOX 2: The DC-X and DC-XA Programs

In 1990, the Ballistic Missile Defense Organization (BMDO) was exploring concepts to launch hundreds of space-based interceptors designed to provide the nation with a shield against incoming ballistic missiles. In the view of BMDO officials, a fully reusable single-stage-to-orbit (SSTO) vehicle was the most promising concept for reducing space transportation costs.

BMDO initiated the Single Stage Rocket Technology program to design an SSTO vehicle, then fly a subscale experimental vehicle as a proof-of-concept demonstrator. Four contractors were selected to develop SSTO designs between 1990 and 1991. Primarily because of funding limitations, McDonnell Douglas, which proposed a vertical-takeoff/vertical-landing configuration, was the only firm selected to build and fly the experimental DC-X vehicle. Using support from the U.S. Air Force's Phillips Laboratory, a streamlined management structure at BMDO, and a rapid prototyping team at McDonnell Douglas, the DC-X was built in 24 months and flown for the first time on August 18, 1993.

The DC-X flew two more times in the summer of 1993 before flight testing was terminated for lack of funding. With funding support from NASA, BMDO was able to restart the DC-X test flight program in the summer of 1994. During the second flight of this series (flight number 5), a detonation in the ground support equipment ripped a large hole in the vehicle's composite skin. Despite the damage, the vehicle executed a successful emergency landing on the desert floor and was subsequently returned to the factory for repair. During the repair interval, the DC-X contract and all program management responsibility was transferred from BMDO to the Phillips Laboratory at Kirtland Air Force Base, New Mexico.

Several lessons have been learned from the DC-X program to date. Namely, streamlined government management of experimental programs is capable of reducing development costs while meeting schedule milestones; "aircraft-like" operations (including rapid turnaround between launches) and maintenance of space transportation systems are feasible; and the number of people required to operate a space transportation sytem can be reduced through automated test and control. Furthermore, the DC-X successfully flight tested several critical—but not all encompassing—SSTO technologies such as navigation aided by the Global Positioning System and a gaseous oxygen/hydrogen reaction control system.

The next flight series of the DC-X is scheduled for May 1995 through July 1995. This flight test phase is scheduled to culminate with the DC-X performing a "pitch maneuver" by which the vehicle transitions from the nose-down position required for atmospheric reentry, to the tail-first position required for landing. Upon completion of the DC-X flight test program, Phillips Laboratory will transfer the vehicle and test equipment to NASA for the DC-XA program. The DC-XA program will incorporate numerous additional critical SSTO technologies into a highly modified DC-X and flight test them in mid-1996. NASA has designated Phillips Laboratory to manage the DC-XA flight test program.

SOURCE: U.S. Air Force, 1995.

ket. Boxes 3 and 4 summarize the CANs for the X-33 and X-34, respectively.

In addition to its RLV efforts, NASA plans to continue its support of ELV development—although in a more reserved manner. NASA's implementation plan states that "in coordination with DOD, NASA will continue to support [ELV] industry initiatives where the use of NASA facilities and resources can minimize the cost and enhance the value of the technology efforts."[29]

Furthermore, NASA's implementation plan states that NASA retains the right to procure new ELV services "where mission-unique modifica-

[29] NASA, "NASA Implementation Plan for the National Space Transportation Policy," Washington, DC, Nov. 7, 1994, p. 25.

BOX 3: The X-33 CAN

NASA issued a cooperative agreement notice for the X-33 reusable launch vehicle on January 12, 1995. This CAN solicited proposals "that offer a new way of doing business, consistent with space policy."[1] The CAN divided the X-33 program into three phases: Phase I—Concept Definition/Design; Phase II—Design/ Demonstration; and Phase III—Commercial RLV Development/Operation. Phase I commenced in March 1995 and is scheduled to run for 15 months. Phase II, which includes design, build, and flight demonstration, will begin by the end of FY 1996 and continue though the end of the decade. At the end of the decade, the government and private sector will jointly decide whether or not to proceed with Phase III.[2]

NASA requested detailed proposals from industry on Phase I and preliminary plans for Phases II and III. Four teams, lead by Lockheed Advanced Development, McDonnell Douglas Aerospace, Rockwell International, and Space Access, Inc., submitted proposals to NASA on February 24. NASA selected three (Lockheed Martin, McDonnell Douglas, and Rockwell) to proceed with Phase I. NASA and the selected companies will share expenses during Phase I, with a total NASA contribution of $24 million.[3]

Selection of one or more X-33 concepts to proceed into Phase II will be based on specific criteria (which NASA reserves the right to change at any time up until Phase II proposals are requested) that reflect mature business and design plans.[4]

The technical composition of the X-33 program is based on the goal of continually lowering the cost of access to space "to promote the creation and delivery of new space services and other activities that will improve economic competitiveness."[5] In the technical description of the X-33, the CAN states that "the X-33 must adequately demonstrate the key design and operation aspects of [a single-stage-to-orbit] RLV rocket system. . . ."[6] The CAN notes that an SSTO rocket system is the *goal* of Phases I and II "because past studies indicate it has the best potential for achieving the lowest cost access to space while acting as an RLV technology driver. . . ."[7] Nevertheless, the CAN grants the private sector the option of proposing any RLV technology—not necessarily SSTO—in Phase III.

The government contribution to the X-33 program for Phases I and II (i.e., FY 1995-99) is projected to be roughly $662 million. During the same period, an additional $339 million has been allocated for core RLV technology development programs.[8] NASA expects industry to fund final development (i.e., Phase III) of a commercial RLV, although the Agency realizes that a small government contribution may be necessary.

---

[1] NASA, "A Cooperative Agreement Notice: Reusable Launch Vehicle (RLV) Advance Technology Demonstrator—X-33," Jan. 12, 1995, p. ii.

[2] Ibid., p. ii-iii.

[3] NASA, "X-33, X-34 Contractors Selected for Negotiations," NASANews Press Release, Mar. 8, 1995.

[4] NASA, op. cit., footnote 1, p. iii.

[5] Ibid., p. A-2.

[6] Ibid.

[7] Ibid.

[8] "Supporting technology demonstration" and "RLV technology program (focused phase I and current NRA activities)" are often collectively referred to as the "core technology development programs." (See table 8.)

SOURCE: Office of Technology Assessment, 1995.

BOX 4: The X-34 CAN

NASA issued a cooperative agreement notice for the X-34 RLV on January 12, 1995. According to the CAN, "the intent of (the X-34) solicitation is to stimulate the joint industry/government funded development of a small reusable, or partially reusable, booster that has potential application to commercial launch vehicle capabilities, which will provide significantly reduced mission costs for placing small payloads in LEO. The booster must demonstrate technologies applicable to future reusable launch vehicles."[1]

The X-34 program is much shorter than the X-33 program and is not divided into major phases. Nevertheless, the CAN does set out three milestones for the X-34 program: test flights beginning in late 1997; orbital launch by mid-1998; and two test flights later in 1998 as a NASA research platform.

After reviewing CAN proposals from Space Access, Kelly Space and Technology, and a team led by Orbital Sciences, NASA selected the Orbital Sciences team (Orbital Sciences and Rockwell International) as the contractor for the X-34.[2] The proposed government contribution to the X-34 program for FY 1995-99 is $70 million. Orbital Sciences and Rockwell plan to invest $50 million each—for a total program cost of $170 million.[3]

Despite losing the X-34 competition, both Space Access and Kelly Space and Technology are reportedly looking for private backing for their concepts. Each company plans to continue its development effort without government support.[4]

[1] NASA, "A Cooperative Agreement Notice: Reusable Launch Vehicle (RLV) Small Reusable Booster—X-34, "Jan. 12, 1995, p. ii.

[2] NASA, "X-33, X-34 Contractors Selected for Negotiations," NASANews Press Release, Mar. 8, 1995.

[3] Any cost overruns on the X-34 program will fall upon the Orbital Sciences/Rockwell team. See Ben Iannotta, "OSC, Rockwell Selected To Run X-34 Project," *Space News*, Mar. 13, 1995, pp. 4, 37.

[4] Ibid.

SOURCE: Office of Technology Assessment, 1995.

tions are required of the existing medium/heavy-lift vehicles."[30] This statement signals NASA's intent to proceed with a special procurement of a new, low-end medium or "Med-Lite" launch vehicle, and according to the implementation plan, "NASA's budget contains funding to continue to acquire launch services from the U.S. commercial ELV industry to support civil government launch service requirements."[31]

X-33

The X-33 Advanced Technology Demonstrator is the flagship of NASA's RLV Technology Program. NASA hopes that the X-33 will provide the springboard to a commercial RLV in the medium-to-heavy payload class that radically reduces the cost of access to space while improving both reliability and operability.

The CAN does not restrict the X-33's commercial follow-on to any one particular RLV concept, and NASA officials insist industry will decide for itself what kind of commercial RLV to build. Nevertheless, NASA believes that an SSTO space transportation system—if it proves technologically feasible—will be the lowest cost solution (see box 5). Therefore, NASA has limited X-33 development to the SSTO concept.

[30] Ibid.

[31] Ibid.

BOX 5: Single Stage To Orbit

Since at least the early 1960s, launch vehicle engineers have dreamed of building reusable launchers because they offer the potential of relative operational simplicity and reduced costs compared with expendable vehicles.[1] Until recently, the necessary technologies were not available. Now, thanks to recent technology improvements,[2] many engineers believe that the United States is technically able to design and produce an SSTO vehicle with sufficient payload capacity to meet most government and commercial space transportation requirements.

Notwithstanding these recent technological advances, the development of an SSTO space transportation system involves significant risk. For example, because an SSTO launch vehicle will have no expendable components, it will need to carry more fuel than would otherwise be necessary if it were shedding weight by dropping stages during ascent. Achieving the fuel mass fraction[3] necessary to reach orbit with a useful payload will require a host of technological advances that improve fuel efficiency and lower structural weight without compromising structural integrity. Additionally, completely reusable launch vehicles are technologically much more difficult to achieve because components must be capable of resisting deterioration and surviving multiple launches and reentries.

The best configuration—if there is one—for an SSTO vehicle is yet undetermined. The difficulties of a horizontal takeoff-horizontal landing vehicle are known from the National Aerospace Plane Program, and it is likely that this configuration will not receive much consideration. Three other potential configurations pose unique technical obstacles, but are likely to receive more serious consideration: vertical-takeoff/vertical-landing, vertical-takeoff/horizontal-landing (winged body), and vertical-takeoff/horizontal-landing (lifting body).

Vertical-takeoff/ vertical-landing

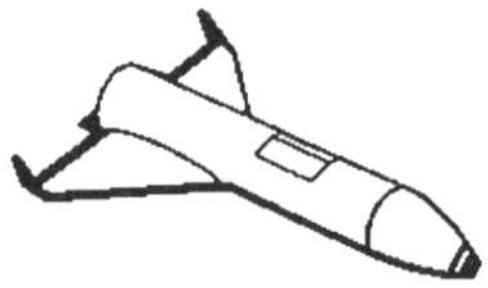

Vertical-takeoff/ horizontal-landing (winged body)

Vertical-takeoff/ horizontal-landing (lifting body)

The decision to focus on SSTO development in the X-33 program was based on NASA's 1994 "Access to Space Study." This study investigated three options for reducing the cost of access to space. The first option called for an overhaul of several Space Shuttle subsystems and continued operation of an upgraded Space Shuttle. NASA estimated that such an operation would fail to reduce sufficiently the cost of access to space. The second option proposed the development of new multistage ELV technologies that would deploy either a cargo pod or manned vehicle. While NASA viewed this option as more appealing than the first, it was less attractive to NASA than the third option—development of an SSTO.

[1] For the early history of attempts to build reusable launch vehicles, see Richard P. Hallion (ed.), *The Hypersonic Revolution: Eight Case Studies in the History of Hypersonic Technology*, Volume II (Wright-Patterson Air Force Base, Ohio: Special Staff Office, Aeronautical Systems Division, 1987), p. 948.

[2] Some of these improvements were funded through the National Aerospace Plane Program (1986-94). The goal of NASP was to design and build a horizontal takeoff-horizontal landing, air-breathing SSTO.

[3] The term *fuel mass fraction* refers to the ratio of the weight of fuel required to accomplish the mission to the total initial vehicle weight (excluding payload). An SSTO launch vehicle will require a higher fuel mass fraction (i.e., lower structural weight fraction) than existing multiple-stage vehicles in order to reach orbit.

SOURCE: Office of Technology Assessment, 1995.

Pending the successful completion of the X-33 program, NASA sees the RLV as a potential replacement vehicle for the Space Shuttle.[32] For this reason, the X-33 CAN requires that the commercial follow-on RLV support the International Space Station. The X-33 CAN sets the required dimensions of the follow-on RLV's payload bay to a 15 ft diameter and a 30 ft length, with a required mass payload deployment capability of 20,000 to 25,000 lbs to a 220 nmi (maximum 244 nmi) altitude inclined at 51.6°.[33]

Although the CAN requires a commercial follow-on to the X-33 to service the International Space Station, it requires only suborbital flight of the X-33 during Phase II test flights. Table 7 summarizes the technical requirements for the X-33 program.

The CAN outlines the cost-sharing arrangement between government and industry for the development and operation of the X-33. Table 8 shows NASA's projected funding contribution for Phases I and II. NASA expects that industry will roughly match the government's overall Phase I/II contribution. A number of core technology development programs were started in FY 1994,[34] and Phase I award winners have already been selected. Table 9 summarizes the Phase I award winners, their major team members, and their respective vehicle concepts.

Although NASA has the lead role in RLV development, DOD maintains a supporting role.[35] Within DOD there are both skeptics and optimists about whether RLV technology will mature to a point where it will be useful to the Department. Nevertheless, DOD engineers and managers hope to retain some influence over the direction of RLV development so that DOD will be in a position to benefit from future RLVs.

Already DOD has budgeted $20 to $30 million per year for RLV-related technology projects over the projected life of NASA's program.[36] Additionally, DOD has accelerated some of its propulsion technology programs to provide data to NASA to support its December 1996 decision on the X-33. DOD is also involved in materials research that could be useful to RLV development, and it will play a key role in helping NASA design and execute a launch operations concept for both the X-33 and X-34 programs.[37]

X-34

NASA views the X-34 program as a means of gaining early experience with government-indus-

---

[32] Industry may choose to develop one RLV to capture NASA flights and another to serve the commercial space transportation market.

[33] Text in the CAN states that this requirement "is a preliminary assessment of minimum single payload element weights that are required to support the Space Station [and that this] estimate is for planning purposes only and [does] not represent a commitment by the Space Station Program." See NASA, "A Cooperative Agreement Notice: Reusable Launch Vehicle (RLV) Advance Technology Demonstrator—X-33," Washington, DC, Jan. 12, 1995, p. vi. NASA officials contend that this language allows for the possibility of satisfying the 20,000- to 25,000-lb payload delivery requirement with multiple flights and that the ultimate decision on RLV payload capacity will be made by industry. Industry officials and other analysts note, however, that NASA is not precluded by this text from setting a final requirement that dictates a payload delivery capability of 20,000 to 25,000 lbs to International Space Station orbit in a single flight.

[34] The $20 million budgeted in FY 1995 for these core technology development programs is controlled by DOD. The DOD Comptrollers Office has held up distribution of this money to Phillips Laboratory. Phillips Laboratory, once it receives the money, is expected to use it to support the ongoing core technology development programs. The delay in transferring this money could potentially jeopardize successful completion of Phase I by each of the remaining contractors.

[35] The White House Office, Office of Science and Technology Policy, op. cit., footnote 3, sec. III(2)(c).

[36] U.S. Department of Defense, personal communication, February 1995.

[37] NASA and the U.S. Air Force's Phillips Laboratory are taking steps to ensure technical coordination between DOD and NASA on RLV-related technology development.

TABLE 7: Minimum X-33 and Corresponding RLV Requirements

	X-33	Corresponding RLV
CAPABLE		
Performance		
Sub-orbital, reusable rocket-based flight system	REQ	N/A
Mission applications:	N/A	REQ
Payload delivery: government (civil/military) and commercial missions.		
Capable of delivering/returning cargo and crew complement to/from the International Space Station (ISS) in accordance with ISS requirements (e.g., minimum sizes, loads, schedule).		
▪ ISS located at 220 nmi (244 nmi max.) altitude and 51.6 inclination.		
▪ Current estimated payload delivery requirement: 20,000-25,000 lbs.		
Launch and Flight Operations		
Automated pre-flight and flight operations (launch, ascent, on-orbit, reentry, landing).	REQ	REQ
The flight vehicle shall be capable of safely aborting to the launch site during the ascent phase if required.	GOAL	REQ
7-day maximum mission duration.	N/A	REQ
7-day ground processing time from landing to launch.	GOAL	GOAL
3.5-day ground processing time from landing to launch for reflight under emergency conditions.	GOAL	GOAL
On-Orbit Operations		
The system shall be able to autonomously rendezvous and station keep with the ISS and other orbital spacecraft.	N/A	AS REQ
The system shall be able to autonomously dock payloads with the ISS.	N/A	REQ
Accommodate Payloads		
The flight vehicle shall provide standardized structural, mechanical, electrical, communications, and other interfaces to payload.	N/A	REQ
15-ft-diameter x 30-ft-long cargo bay.	N/A	REQ
OPERABLE		
Schedule Dependability		
The probability of launching within TBD days of scheduled is 0.95.	GOAL	REQ
Responsive		
Maximize robustness to adverse weather conditions.	REQ	REQ
Supportable		
Launch and landing at same location (nominal condition).	REQ	REQ
The flight vehicle shall be capable of unplanned landing at alternate landing sites with minimal support equipment/facilities, e.g.:	GOAL	REQ
No existing cryogenic facilities, launch stands/equipment, etc.		
Self-ferry of flight vehicle between landing and launch sites (add-on engines, landing/nav. lights, etc., equipment allowed).		
Maintainable		
To the extent practical, on-board subsystems required for the flight vehicle shall be field repairable/replaceable.	GOAL	REQ
Equipment required to repair, process, and return vehicle to launch site shall be transportable.	GOAL	REQ
RELIABLE		
0.995 probability of safe recovery of the flight vehicle per mission.	N/A	REQ
0.999 probability of safe recovery of the human passengers per mission.	N/A	REQ

NOTE: GOAL=Desirable Attribute; N/A=Not Applicable; REQ=Required; AS REQ=As Required

SOURCE: NASA, "A Cooperative Agreement Notice: Reusable Launch Vehicle (RLV) Advance Technology Demonstrator—X-33," Jan. 12, 1995, p. A-4.

TABLE 8: Proposed NASA X-33 Funding[a]

Phase	I	I & II		II	
	FY 95[b] ($M)	FY 96 ($M)	FY 97 ($M)	FY 98 ($M)	FY 99 ($M)
Concept definition/design (phase I)	18.0	6.0	—	—	—
Design/demonstration (phase II)	—	43.0	147.0	270.0	178.0
Supporting technology demonstration	20.0	—	—	40.0	147.0
RLV technology program (focused phase I & current NRA activity)	51.2	50.7	30.6	—	—
System engineering and analysis	4.6	4.7	4.7	4.7	4.7
Long-term, high-payoff technology investment	8.7	8.6	15.6	25.7	30.1
Total	**102.5**	**113.0**	**197.9**	**340.4**	**359.8**

[a] FY 1995 dollars.

[b] Includes DOD funding.

SOURCE: NASA, "A Cooperative Agreement Notice: Reusable Launch Vehicle (RLV) Advance Technology Demonstrator—X-33," Jan. 12, 1995, p. A-9.

TABLE 9: X-33 CAN Phase I Award Winners

Prime Contractor	Major team members	Vehicle concept	Concept details
Lockheed Martin (Advanced Development, a.k.a. Skunk Works)	Lockheed Martin (5 divisions plus Rocketdyne), Rohr, Allied Signal, Bankers Trust, and Space Express	Vertical-takeoff/ horizontal-landing (lifting body)	126-ft-long, 1.6 million lbs at liftoff (87.5 % fuel)
McDonnell Douglas Aerospace	Boeing, Douglas Aircraft, Rocketdyne, Aerojet, Pratt and Whitney, and Honeywell	Vertical-takeoff/ vertical-landing or vertical-takeoff/ horizontal-landing	To be announced in June 1995
Rockwell (Space Systems Division)	Rockwell (North American Aircraft, Rocketdyne), Northrop Grumman, Federal Express, and Orbital Sciences	Vertical-takeoff/ horizontal-landing (delta wing with twin tails)	X-33: 52 % scale of RLV, 100-ft-long, 55-ft wingspan, 350,000 lbs at liftoff, 3 engines RLV: 187-ft-long, 1.9 million lbs at liftoff, 15,000-40,000 lbs payload

SOURCES: Bruce A. Smith, "NASA Speeds Selection of X-33, X-34 Plans," *Aviation Week and Space Technology*, vol. 142, No. 11, Mar. 13, 1995, pp. 107, 109. Office of Technology Assessment, 1995.

ORBITAL SCIENCES

The partially reusable X-34 hopes to achieve threefold cost reductions for launching small payloads.

TABLE 10: Proposed NASA X-34 Funding[a]

Fiscal year	Funding ($M)
1995	10
1996	30
1997	15
1998	10
1999	5
TOTAL	**70**

[a] FY 1995 dollars.

SOURCE: NASA, "A Cooperative Agreement Notice: Reusable Launch Vehicle (RLV) Small Reusable Booster—X-34," Jan. 12, 1995, p. A-5.

try cooperation and hopes that the program will provide opportunities to develop and demonstrate technologies applicable to future development of a commercial follow-on to the X-33.[38] A low-cost commercial version of the X-34 could significantly reduce NASA's cost for launching small payloads (about 10 to 12 per year) as well as further expand the commercial LEO market for small payloads.

The X-34 CAN sets out three milestones for the program: 1) test flights beginning in late 1997, 2) orbital launch by mid-1998, and 3) use as a NASA test bed later in 1998. The X-34 planned by the team of Orbital Sciences and Rockwell is a partially reusable, two-stages-to-orbit (TSTO) vehicle.[39] NASA's portion of the budget for the X-34 program is provided in table 10. Orbital Sciences and Rockwell will each contribute an additional $50 million and cover any cost overruns experienced by the program.

Space Shuttle

The NSTP also directs NASA to "continue to maintain the capability to operate the Space

[38] In particular, NASA hopes to advance the use of graphite composites in the primary vehicle structure, gain experience with autonomous vehicle health management and monitoring, improve reusable cryogenic tank systems and thermal protection systems, and learn more about RLV operations.

[39] The proposed X-34 will be carried on a large jet aircraft—much like Orbital Sciences' Pegasus SLV—to a specific altitude and distance from the landing field. The aerodynamic, fully reusable booster vehicle of the X-34 will then be launched from the jet, climb out of the Earth's atmosphere, release a non-aerodynamic, expendable orbiting vehicle that is attached to the payload, and then return to its landing field. The orbiting vehicle, after separation from the reusable booster, will continue on and deliver the payload to its intended orbit.

FIGURE 1: Shuttle Phasing Strategy Timeline

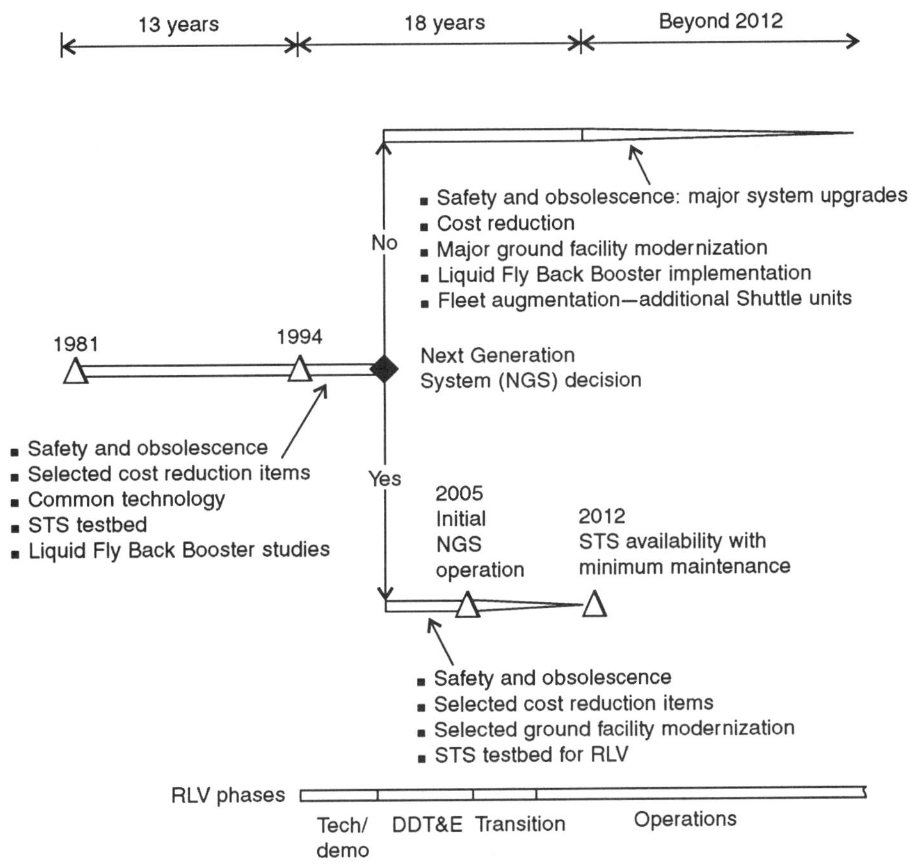

SOURCE: NASA, "NASA Implementation Plan for the National Space Transportation Policy," Washington, DC, Nov. 7, 1994, p. 19.

Shuttle fleet and associated facilities."[40] In doing so, the NSTP notes that NASA should focus on improving "reliability, safety, and cost-effectiveness."[41]

In response to this directive, NASA proposes a phased approach for the Space Shuttle program (see figure 1). Until 2000, the RLV technology development program will proceed on a path to replace the Space Shuttle in 2012. Simultaneously, NASA will continue to upgrade Space Shuttle components and subsystems in an effort to reduce costs and improve safety, performance, and reliability.

NASA contends that in 2000 the government and industry must decide on whether or not to pursue commercial development of a follow-on RLV. If either the government or industry decides to forgo commercial development of an RLV, NASA states that it "may need to embark on a substantial [Space Shuttle] upgrade program" that would in-

[40] The White House Office, Office of Science and Technology Policy, op. cit., footnote 3, sec. III(1).

[41] Ibid., sec. I(2).

clude ground facility modernization, Liquid Fly Back Booster (LFBB) development, and likely additional Shuttles.[42] All improvements will be aimed at reducing cost without comprising Shuttle safety and reliability.

If, on the other hand, the government and industry decide to develop a commercial follow-on RLV between 2000 and 2012, NASA will maintain those activities needed to "ensure that the Shuttle system flies safely, reliably, and at a lower cost until a replacement vehicle is operational."[43] Furthermore, NASA states that many of the technologies developed during the X-33 and X-34 programs could potentially be implemented in the Shuttle system to achieve safety, reliability, and cost objectives.[44]

Med-Lite

NASA issued a request for proposal (RFP) for a new ELV called Med-Lite on December 5, 1994. NASA designed its Med-Lite procurement to meet its planetary mission requirements for the next 10 years. NASA believes the Med-Lite vehicle will fill a launch capacity gap between current small and medium launchers. NASA also believes that procurement of the Med-Lite vehicle will cost less than purchasing existing ELVs for its future planetary missions. NASA's projects a cost of $25 million to $30 million per launch for the Med-Lite.[45]

NASA received two proposals for the Med-Lite procurement on February 28, 1995. One proposal was made by McDonnell Douglas with Orbital Science as a primary partner. The other proposal was made by Russian-owned Polyot, which proposed using the Kosmos launch vehicle for NASA's planetary missions.[46] NASA has officially selected the McDonnell Douglas/Orbital Sciences team for negotiations leading to eventual award of the Med-Lite contract.[47] According to NASA officials, final selection of a Med-Lite contractor will be made by August 1995.[48]

ELV Technology Programs

NASA has a rich history of ELV technology development. In response to the NSTP, however, NASA is reducing its allocations for continued investment in ELV technologies. NASA expects to complete its Cooperative ELV Tasks in 1996 at a cost of $34 million in FY 1996.[49] Nevertheless, NASA will continue to make its facilities and expertise available to DOD to support ongoing efforts to improve existing ELV technology.

Issues for Congress

The NSTP and the DOD and NASA implementation plans raise several issues for Congress relating to the development of new space transportation systems. In particular, this section discusses:

- questions raised by divided development responsibility between DOD and NASA;
- existing impediments to improved interagency coordination;
- the potential for conflicts and redundancies among the development programs;
- the effect of DOD's emphasis on HLVs in the EELV program;

---

[42] NASA, op. cit., footnote 29, pp. 18, 21-22.

[43] Ibid., p. 20.

[44] Ibid., pp. 17-18.

[45] This does not include extra costs that might be incurred due to launch delays. NASA, personal communication, February 1995.

[46] The Med-Lite RFP contained two qualification criteria. First, the prime contractor must be a U.S.-owned company. Second, more than 50 percent of the Med-Lite vehicle must be produced in the United States. The Polyot proposal does not meet either of these qualification criteria.

[47] Warren Ferster, "NASA Makes Med-Lite Award," *Space News*, Mar. 27, 1995, pp. 1, 20.

[48] NASA, personal communication, March 1995.

[49] NASA, Office of Space Access and Technology, "Science, Aeronautics and Technology Fiscal Year 1996 Estimates," Washington, DC, p. SAT 5-5.

- issues surrounding RLV development;
- questions about NASA's current focus on SSTO and what might happen if commercialization of an SSTO is forgone in 2000; and finally,
- NASA's plans and contingencies for the Space Shuttle beyond 2000.

ISSUE 1a: *Divided responsibility and inter-agency coordination*

The NSTP divides the government's primary responsibilities for space transportation between DOD and NASA, but it does not discuss how conflicts in space transportation policy will be resolved between the two organizations. The lack of any such discussion in the policy and OTA's interviews with Administration officials suggest that conflicts in space transportation policy will be resolved on a case-by-case basis via negotiations between DOD and NASA, possibly with some mediation by a third party within the executive branch.[50]

At a minimum, divided responsibility will increase the need for DOD and NASA to coordinate with one another as well as with the private sector, especially if national space policy objectives are to be achieved within tight budget constraints. DOD will have to consult with each party that uses ELV assets in order to manage those assets in a manner conducive to all interests, and the same applies to NASA for RLVs.

That DOD and NASA will adequately account for the interests of all parties is not a certainty, especially as funds available for space transportation diminish. When conflicts arise over how to approach development of new space transportation systems, negotiations may succeed in keeping both DOD and NASA satisfied, but could fail to account for the interests of all relevant parties, especially those in the private sector.

Such negotiations could also lead to programmatic redundancies.[51] The absence of central authority or leadership may allow DOD and NASA to discount potential redundancies and promote those projects that best address their own organizational requirements. As a result, hard space transportation decisions may go unmade.

The imposition of a central authority has been proposed by many analysts and policymakers as a way to better account for all interests and avoid programmatic redundancies.[52] It is not clear, however, that a central authority would necessarily remedy these problems. Both DOD and NASA possess a considerable amount of bureaucratic and political weight. Therefore, competing organizational interests could potentially override the wishes of a central authority.

Existing legal and organizational obstacles may also stand in the way of achieving the level of interagency and private sector coordination sought by a central authority. The recent controversy over NASA's Med-Lite procurement may be emblematic of this problem (see box 6). Med-Lite has engendered a great amount of debate between DOD and NASA, and illustrates how interagency coordination can be precluded by current law, divergent interpretations of that law, and competing organizational interests.

Therefore, although the Administration's policy calls on DOD and NASA to "combine their [ELV] requirements into single procurements when such procurements would result in cost savings or are otherwise advantageous to the government," achieving this level of interagency coordination may prove extremely difficult. Congress may wish to consider legislative action that would facilitate improved interagency coordination on all space transportation policy matters. If Congress decides to take up this issue, a comprehensive review of existing laws affecting such in-

---

50 Potential third parties within the executive branch include OSTP, NSTC, or either the President or Vice President.

51 See the discussion below about potential conflicts and redundancies within the proposed development programs.

52 For example, in the Bush Administration Vice President Quayle was given considerable authority over space transportation policy.

BOX 6: The Med-Lite Controversy

In NASA's view, the Med-Lite procurement will fill a gap in ELV launch capacity that lies between small- and medium-size launch vehicles. NASA officials see the Med-Lite program as a procurement of launch services for planetary exploration that complies with the Launch Services Purchase Act of 1990.

DOD, on the other hand, has several concerns about the Med-Lite program.[1] In particular, DOD officials view Med-Lite as a launch vehicle development program—not a procurement of launch services. In their eyes, the Med-Lite vehicle may threaten the ability of the EELV program to achieve maximum launch cost reductions for DOD.[2] NASA officials reject the claim that Med-Lite is a development program. They contend that DOD's position on the matter is heavily influenced by a traditional DOD philosophy of procuring launch *vehicles*. NASA contends that it procures launch *services*—not launch *vehicles*—and that NASA gets its money back if a launch is canceled.[3]

In response, DOD officials accuse NASA of using this argument to divert attention away from the development nature of the Med-Lite program. DOD points to the unusually long procurement period of four years[4] for the Med-Lite as an indication that, indeed, the procurement is a development program. Furthermore, DOD officials claim that the Med-Lite request for proposal clearly lays out a development program.[5]

This dispute over procurement philosophy shows just how difficult it is to reconcile conflicting interests. This difficulty becomes even more apparent when one reviews the divergent interpretations regarding the legality of the Med-Lite procurement. DOD has suggested that the Med-Lite procurement may be in violation of Public Law 102-139, Title III, October 28, 1991, section 2459d entitled "Prohibition of grant or contract providing guaranteed customer base for new commercial space hardware or services." This statute

[1] DOD, "Med-Lite: A DOD White Paper," Washington, DC, December 1994.

[2] This point is discussed in further detail later in this report.

[3] DOD officials take issue with this characterization of DOD launch procurement by NASA. In their view, the Department procures launch *services* as well, but writes in reflight provisions in its contracts in the event that a launch is canceled.

[4] DOD notes that "typical lead times for procurements conducted by commercial satellite builders are about two years [and that] NASA typically orders expendable launch vehicles 24-30 months ahead of launch." See DOD, op. cit., footnote 1, p. 3.

[5] DOD originally made this observation based on a draft Med-Lite request for proposal dated September 27, 1994, and noted that this draft RFP provided for "extensive NASA oversight through design review, technical interchange meetings, hardware acceptance reviews, quarterly program reviews, daily engineering interface, independent flight assurance reviews, hardware and software pedigree reviews, full approval of designs and modifications, and a series of mission and launch readiness reviews." See DOD, op. cit., footnote 1, p. 3.

teragency coordination may be a prudent first step, as there will likely be much disagreement on how the necessary reforms should be formulated.

ISSUE 1b: *Potential conflicts and redundancies*

NASA has a number of space transportation programs underway. Work has already begun on both the X-33 and X-34 programs; test flights of the DC-XA resumed in May 1995; upgrades to the Shuttle continue; and the Med-Lite procurement is moving forward. DOD, for its part, is pressing ahead with the EELV program and at the same time continuing some upgrades to the existing ELV fleet. All of this amounts to a sizable portfolio of new space transportation technology development and procurement. While this multitrack approach may reduce the overall risk of pursuing new space transportation systems, it may also lead to potential conflicts and redundancies.

EELV and RLV

NASA officials hope the proposed X-33 development program will lead to a commercial RLV that will provide dramatically lower launch costs than

BOX 6 (cont'd.): The Med-Lite Controversy

states that "no amount appropriated to the National Aeronautics and Space Administration in this or any other Act with respect to any fiscal year may be used to fund grants, contracts or other agreements with an expected duration of more than one year, when a primary effect of the grant, contract, or agreement is to provide a guaranteed customer base for or establish an anchor tenancy in new commercial space hardware or services. . ."[6] DOD officials think that NASA has designated its planetary missions as an anchor tenant for Med-Lite and notes that it is unusual for any government entity to purchase launch services for yet unspecified payloads.[7]

NASA officials counter this contention and argue that, in fact, the Med-Lite procurement is required by the law. Pointing to the Launch Services Procurement Act of 1990, NASA officials believe that—with minor exception—it must purchase launch services in a competitive manner from commercial providers for its primary payloads.[8] NASA officials do not believe that any of the exceptions provided for in the law apply. Therefore, in their view, NASA must go forward with the Med-Lite procurement.

While divergent legal interpretations have proven problematic, other legal intricacies not under dispute have also influenced the Med-Lite debate. DOD offered to launch three NASA planetary exploration missions on three Titan II launch vehicles. In return, DOD asked only that NASA pay for refurbishment and launch operations at an approximate cost of $18 million per vehicle. This option, however, was legally precluded by the Economy Act, which requires that DOD sell its launch vehicles to NASA at full price—approximately $54 million per vehicle. This difference in cost effectively precluded NASA from considering this option because the Med-Lite cost target is $25 million to $30 million per launch.

---

6 DOD, op. cit., footnote 1, p. 3-4.

7 NASA projects five "firm" launches between now and 1999. Only three, however, have been named: Mars Orbiter-2, Mars Lander-1, and the Far Ultraviolet Spectroscopic Explorer (FUSE). See Warren Ferster, "NASA Makes Med-Lite Award," *Space News*, Mar. 27, 1995, pp. 1, 20.

8 Exceptions include instances in which "(1) the payload requires the unique capabilities of the space shuttle; (2) cost effective commercial launch services to meet specific mission requirements are not reasonably available and would not be available when required; (3) the use of commercial launch services poses an unacceptable risk of loss of a unique scientific opportunity; or (4) the payload serves national security or foreign policy purposes." See Title II, Section 204b of the National Aeronautics and Space Administration Authorization Act, FY 1991.

SOURCE: Office of Technology Assessment, 1995.

existing launch systems. DOD's investment in EELV could potentially preclude or hinder achievement of this objective. If the EELV program succeeds in making the U.S. space transportation industry more competitive in the global marketplace for launch services, the incentive to sustain a continuous stream of private investment into the commercial development of an RLV over several years may well diminish.

Nonetheless if commercial development of an extremely low-cost RLV proceeds, then, at a minimum, the RLV will compete with the EELV for payloads. This competition could prevent the EELV program from achieving long-term cost reduction targets set by DOD. For example, if extremely low launch prices were charged during the pioneering flight stage of the RLV, the RLV would probably attract payloads away from the EELV. This would reduce EELV production volumes—offsetting or potentially outweighing any gains in production volume created by commonality within the EELV family.

This conflict between the EELV and RLV programs has prompted some analysts to question the prudence of pursuing both programs simultaneously. Some have suggested forgoing the EELV

ROCKWELL INTERNATIONAL

Artist's conception of Rockwell's proposed commercial follow-on RLV to the X-33.

program and investing the $2 billion in RLV development instead. They argue that the existing ELV fleet can adequately support DOD's manifest of payloads indefinitely or until a low-cost RLV is developed.

In response to this proposal, DOD officials note that, in an environment of declining space budgets, the Department must act now to lower its launch costs. They further note that there is substantial uncertainty surrounding the RLV development program and its ability to achieve radical launch cost reductions. Therefore, DOD officials—who are fairly confident that the EELV program can reduce overall launch costs for the Department—believe they have chosen a prudent course of action.

NASA officials offer a similar line of reasoning for the RLV development program to those that suggest it be eliminated in favor of a scaled-up EELV program. Current operating costs for the Space Shuttle and growing budget constraints have put NASA in the position of pursuing SSTO as a potentially low-cost Shuttle alternative, and NASA officials argue that any delay in pursuing SSTO would require a major investment in the aging Shuttle fleet to keep it in operation beyond 2012.

X-33 and X-34

NASA officials issued CANs for both the X-33 and X-34 because they believe that a successful technology development and demonstration effort must fund a diverse number of projects. While conceding that some projects will be successful in maturing the technology and others will not, they argue that a premature cessation of any given project would leave the overall program vulnerable to reliance on a potentially "wrong" technology.

A dual-track strategy, however, is most effective when both tracks are on course to solve the same problem. The X-33 and X-34 programs do not address the same problem. The X-33 program is focused on developing a fully reusable SSTO to replace the Space Shuttle (i.e., for use as a medium-to-heavy-lift booster). The X-34, on the other hand, addresses the problem of developing a partially reusable launch system for delivering small payloads to orbit (i.e., a small-sized booster). Therefore, critics of NASA's dual-track strategy contend that the absence of one or the other X-vehicle would not increase the likelihood of arriving at the "wrong" technological answer, because each program—from the start—has been designed to pursue a different technological answer.

Some analysts and policymakers have suggested canceling the X-34 program based on this argument. Others have suggested terminating the X-34 program because they believe it is more of an operational vehicle development program than a true experimental vehicle program. Those making this latter argument note that the X-34 program is scheduled to conduct only a small number of test flights (which critics do not believe will be conducted in time to inform X-33 development), with only two technology demonstration flights before becoming a commercially operated vehicle sometime in 1999.[53]

Despite the criticisms leveled against the X-34 program, there are several reasons for proceeding with X-34 development. First, the X-34 could potentially provide an early in-flight test bed for RLV-related technologies.[54] This experience could positively affect the design of the X-33 by steering it toward or away from certain technologies. Second, one of the objectives of the X-34 program is to achieve a threefold reduction in the cost of access to space for small payloads.

NASA and other government payloads constitute a major portion of the small payload market.[55] Therefore, if the X-34 program succeeds in achieving targeted cost reductions, NASA—for a relatively small investment in the X-34 program—will have achieved significant long-term savings for the government and will quickly recoup its $70 million investment.[56] Additionally, if the X-34 dramatically reduces the price charged for launching small payloads, the commercial benefit to the United States could show up in a larger market share of the global launch services industry, expanded space-based business opportunities, and lower prices for consumers for both existing and new space-based services (e.g., telecommunications).

Med-Lite and EELV

As noted in box 6, NASA is committed to the Med-Lite procurement while DOD officials are concerned that the Med-Lite vehicle may threaten the ability of the EELV program to achieve maximum launch cost reductions for DOD. DOD officials would be less concerned if they felt that the manufacturer of the Med-Lite could successfully scale up its vehicle design to meet DOD heavy-lift requirements. If this were possible, then the Med-Lite would be a potential EELV candidate. But because DOD officials have little confidence that the Med-Lite can be successfully scaled up, they are concerned that the Med-Lite might eventually compete with the EELV for medium payloads—thereby reducing EELV production volumes and possibly dampening overall EELV cost savings.[57] Additionally, the Med-Lite vehicle may slow, if not undermine, the industry consolidation currently underway in space launch services that some DOD officials believe is necessary.

Space Shuttle and RLV

NASA has proposed to phase in any newly developed RLV between 2005 and 2012. During this period, the Space Shuttle would continue flying, while the RLV would fly only a few times a year. The yet unproved reliability of the RLV combined

---

[53] At present, NASA has yet to decide what technologies it wishes to test on these flights. NASA officials are contemplating using the flights to conduct high-speed aerodynamic, aeropropulsion, or structural tests. Although such tests would be conducted on the reusable suborbital rocket booster, they would not necessarily be geared toward testing reusable launch technologies. Because the small reusable rocket booster resembles an aircraft in many respects, the tests could potentially be geared to test aircraft-related technologies instead.

[54] Many RLV-related technologies have undergone extensive ground testing. This testing seldom mimics what might occur in flight. Therefore, first-time flight testing of these technologies on a relatively inexpensive X-34 may be more prudent than trying them out for the first time on either the Space Shuttle or the X-33.

[55] Recall that NASA plans to launch 10 to 12 small payloads per year over the next 10 years. DOD will probably contribute another 1 to 4 small payloads per year over the same period of time.

[56] If the X-34 achieves a threefold reduction in the cost of access to space for small payloads, NASA would recoup its investment after launching approximately seven payloads of an average weight of 1,500 lbs. This estimate would vary depending on the actual cost reductions achieved as well as the actual weights of the payloads launched.

[57] Recent comments by industry officials regarding Med-Lite indicate that the threat perceived by DOD may not be very real. Industry seems to be skeptical of the long-term viability of a Med-Lite vehicle—instead viewing the EELV as the vehicle of the future. See Warren Ferster, op. cit., footnote 47.

with the need to fly missions to the International Space Station would probably preclude the option of grounding the Space Shuttle during this period. Therefore, NASA might require substantial funding during these years in order to support the simultaneous operation of the Space Shuttle and flight testing of the RLV.

ISSUE 1c: *HLVs drive the EELV program*

Historically, cost has taken a back seat to mission in defense-related space activities. Budget reductions, however, have prompted DOD to reduce how much it spends on space transportation. The U.S. Air Force plans to reduce its out-year space transportation budgets by downsizing its payloads and phasing out its heavy-lift requirements.[58] The intelligence community, however, has a continued need for HLVs.[59] Therefore, because DOD believes larger cost savings are possible in the HLV class than in the MLV class, it has geared the EELV program toward achieving significant HLV cost reductions.[60]

Some industry officials have expressed concern that DOD's focus on HLV cost reductions ignores private sector concerns.[61] In their view, the market for medium payloads is the biggest portion of the satellite market. Therefore, U.S. launch providers, satellite owners, and ultimately the consumers of space-based services would much prefer cheaper, more reliable MLVs over cheaper, more reliable HLVs. Others note that commercial geosynchronous-Earth-orbit (GEO) payloads are getting heavier and contend that significant HLV cost reductions would be commercially attractive,

LOCKHEED MARTIN

DOD has a continuing need for HLVs, such as this Titan IV. A heavy-lift EELV is scheduled to be operational in 2005.

as evidenced by ESA's development of the heavy-lift Ariane 5. It is unlikely, however, that the heavy-lift EELV will be able to compete successfully in the commercial market for launch services—even if it achieves a 40-percent cost reduction over the Titan IV.[62]

---

[58] Currently, the U.S. Air Force is purchasing enough Titan IVs to launch its remaining heavy payload manifest.

[59] U.S. Department of Defense, personal communication, January 1995. See also U.S. Department of Defense, op. cit., footnote 15, p. II-11.

[60] DOD officials expect that the EELV program will achieve a maximum cost savings of 40 percent for HLVs, while only achieving a maximum of 10 percent cost savings for MLVs. U.S. Department of Defense, personal communication, January 1995.

[61] See the section covering fundamental objective #4 for a more developed discussion of the private sector role in space transportation policy decisionmaking.

[62] ESA developed the heavy-lift Ariane 5 so that Arianespace will be able to continue its practice of launching multiple payloads per launch when such a practice is feasible.

ISSUE 1d: *RLV development*

NASA has pointed to its RLV development strategy as one example of its "new way of doing business." This section examines some of the concerns that have emerged outside of NASA as the result of its proposed RLV development strategy.

Property and Data Rights and Core Technologies

Both NASA and its X-34 industry partners have argued that technical data acquired on the X-34 program could be of potential use to the X-33 program. Even if the technical data gathered on the X-34 program were useful, however, any effort to transfer that data to the X-33 program may face notable difficulties.

The CANs for both the X-33 and X-34 grant specific property and data rights to both prime and lower-tier contractors. For example, the CANs state that NASA—in the event of a joint invention with an industrial partner—will attempt to "refrain from exercising rights which would adversely affect commercialization" by that industrial partner.[63]

NASA's desire to transfer technology from the X-34 to the X-33 may put a strain on its ability to adhere to the spirit of this language. This may be especially true if a contractor or subcontractor that has developed technology for the X-34 is not part of the X-33 development team. It would be less problematic if firms developing technology for the X-34 were also on the X-33 team.[64]

Concerns about property and data rights are not limited to technology transfer from the X-34 to the X-33. In conversations with industry, OTA has learned that property and data rights issues surrounding the proposed RLV core technology development programs are a point of major concern, particularly those core technology development programs planned for Phase II of the X-33 program.[65] At present, NASA has not told industry whether these core technology development programs will be controlled by NASA, awarded competitively, or granted to the X-33 contracting team.[66]

Industry officials contend that if NASA retains control or decides to offer these core technology development programs to all of industry, property and data rights issues could—in one way or another—potentially hamper technology transfer or commercialization.

For example, if NASA decided to honor the X-33 CAN language concerning property and data rights and allow firms to retain title to property and data rights,[67] technology transfer would depend on industry negotiations. These negotiations could potentially slow or thwart the commercialization of a follow-on RLV if compensation paid to firms developing core technologies significantly reduced the return on investment for developers of the RLV. Also, industry negotiations could result in a waste of government investment if developers of the RLV dismiss technologies developed by other firms.

If NASA decides to take the opposite approach and exercise its title rights, it could transfer a technology out of the core technology development programs and into the commercial development of the follow-on RLV without compensating the firm that developed the technology. Any firm losing its title rights might reasonably complain that NASA had violated the spirit of the CAN by

[63] NASA, op. cit., footnote 33, p. ii.

[64] Rockwell International is the only firm competing as a major team member on both X-vehicles. Orbital Sciences, for its part, has told OTA of its willingness to share new X-34 technology with NASA and X-33 contractors.

[65] NASA projects that it will contribute roughly $187 million toward core technology development during Phase II.

[66] NASA's current position appears to be that it will control core technology development program decisions and obtain input pertinent to making those decisions from the X-33 contracting team.

[67] In the case of inventions developed by small or disadvantaged businesses or nonprofit organizations, NASA must, by law, allow such firms to keep their property and data rights if they elect to retain them. See NASA, op. cit., footnote 33, p. iii.

U.S. DEPARTMENT OF DEFENSE

The YF-22 (top) beat the YF-23 (bottom) in a head-to-head fly-off competition for the U.S. Air Force's advanced tactical fighter procurement.

undercutting the firm's ability to commercialize technologies in which it invested.

And finally, if NASA chooses to compete the core technology development programs on the open market, NASA would effectively be the entity to decide what core technologies should or should not be developed, rather than developers of the RLV. While this may not be an unusual role for government, it may not be the most cost-effective path to a commercial RLV, especially if developers of the RLV decide to fund additional core technology programs of their own.

In any case, industry teams preparing business plans for Phase II proposals of the X-33 program will want to have a clearer understanding of how funds for the core technology development programs will be distributed before they settle on their final Phase II proposals.

X-33: Alternative Programmatic Approaches

Some critics of the proposed X-33 program argue that NASA should look back to the early days of jet aircraft for a development model. They argue that the X-33 should not be designed in the shadow cast by future requirements for the International Space Station. Instead, NASA should fully fund an X-33 program that focuses solely on demonstrating SSTO technology.[68]

NASA contends that, although the X-33 program is not fully funded, it is, indeed, an X-program that focuses solely on demonstrating SSTO technology. NASA justifies its approach because of declining space budgets and because the eventual intent of the program is to commercialize a follow-on RLV. Therefore, NASA believes that industry should be expected to contribute to X-33 development.

Others have suggested a competitive fly-off between or among competing X-33 concepts.[69] NASA has expressed an interest in conducting such a fly-off. Proponents of a fly-off strategy believe that it would decrease the possibility of choosing the "wrong" technology and increase the likelihood of retaining competition in the domestic launch vehicle industry.

To conduct a fly-off that would be technologically meaningful, the government would need to budget more money in the near term than currently planned. This increased government investment

[68] Some proponents of this view are suggesting a figure of approximately $2 billion for the period spanning FY 1996-2000, an amount comparable to DOD EELV funding.

[69] A fly-off would entail development of multiple versions of the X-33 by competing teams. Each version of the X-33 would undergo a similar regime of flight tests. At the end of flight testing, each team could move forward with commercialization of its concept if it chose to. A fly-off competition need not be limited to SSTO concepts. Such a fly-off could be open to other RLV concepts such as TSTO.

runs counter to the trend toward smaller government space budgets. Moreover, a fly-off strategy would entail significant financial risk for industry participants. This financial risk might be lowered by a winner-take-all fly-off strategy, with the loser being reimbursed for its efforts. This option, however, would further increase the cost to government for pursuing such a strategy.

The Role of DOD Payloads in RLV Development

Many of the industry-government partnership schemes for RLV funding under debate within the executive branch contain an implicit requirement that some DOD payloads be committed to early RLV launches. DOD officials cite the Space Shuttle experience as their reason for being hesitant to make any such commitment. DOD officials also note that the proposed RLV will initially only place payloads in LEO at altitudes comparable to that of the International Space Station. Therefore, because most DOD payloads are placed in higher orbits, an upper stage would be required for DOD payloads to reach their mission orbit. The addition of an upper stage would add technical complexity and increase the cost for such missions.[70]

Excluding DOD payloads during the pioneering flight phase of the commercial follow-on RLV would drive the price of launching the RLV higher if other payloads cannot be attracted to fly aboard the new system.[71] The commercial RLV developer could attract other payloads by offering to deploy them at little to no cost during RLV test flights. Critics of this proposal note, however, that it would undercut other U.S. commercial launch vehicles and limit their ability to become more competitive by taking away a piece of their market, thereby reducing their volume of production.

At this time it is unclear how this issue will be resolved. NASA and its commercial partners need a sufficient number of payloads to both attract potential investors and prove the reliability of RLV technology during its pioneering flight stage. DOD wishes to safeguard its missions and is not willing to contribute payloads in the pioneering stage of RLV flight. And attempts to attract other payloads to the pioneering flight stage of the RLV may undercut the commercial position of the rest of the U.S. space transportation industry.

ISSUE 1e: *SSTO?*

The prudence of focusing the RLV development program on the SSTO concept is a matter of some debate. Some industry analysts claim that NASA has prematurely committed to SSTO by unnecessarily ruling out other RLV options, particularly a fully reusable TSTO. In their view, the RLV development program may end up being less revolutionary than it otherwise might be if the TSTO option were retained in the short term. Other analysts believe that SSTO is the only alternative that will sufficiently reduce the cost of access to space and claim that NASA is proceeding too cautiously. Table 11 summarizes the advantages and disadvantages of both the SSTO and TSTO reusable concepts.

SSTO or RLV?

The NSTP assigns NASA the responsibility of leading the effort to develop and demonstrate "next generation reusable space transportation systems" and offers SSTO as only one possible technological option.[72] Yet, the policy also instructs NASA to focus research "on technologies to support a decision no later than December 1996 to proceed with a sub-scale flight demon-

[70] Similarly, most medium-sized commercial satellites are stationed in orbits higher than that of the International Space Station. Therefore, they too would require an upper stage—at added expense—in order to reach their desired orbit.

[71] The non-recurring development costs for a commercial RLV system would have to be amortized over a smaller number of launches, thereby raising the cost per launch.

[72] The White House Office, Office of Science and Technology Policy, op. cit., footnote 3, sec. I(3).

TABLE 11: Pros and Cons of the SSTO and TSTO Reusable Concepts

	SSTO	TSTO
Pros	Uses one vehicle instead of two Avoids cost and complexity of staging Requires less complex aerodynamic analysis and design than TSTO Simplifies ground and flight operations compared to TSTO	Reduces technological difficulty of reaching orbit compared to SSTO Allows for use of expendable thrust augmentation (i.e., strap on boosters) Allows designers to build in larger performance margins[a] than SSTO
Cons	Requires high fuel mass fraction (i.e., low structural weight fraction) not achievable with existing technology Requires performance maximization (i.e., small performance margins)[a] Uses engines at all altitudes, although engines are optimized for one altitude[b]	Adds cost and complexity of staging Makes ground and flight operations more costly and complicated than SSTO Requires more complex aerodynamic analysis and design than SSTO Requires reintegration of stages before reflight

[a] The term "performance margin" refers to the difference between the designed and required performance of any given component or subsystem. A space transportation system with small performance margins must operate close to its design limit. A space transportation system with large performance margins does not have to, but can, operate close to its design limit.

[b] Engines could be optimized for multiple altitudes. This, however, would require complex and expensive variable geometry engines.

SOURCES: Lt. Col. John London, U.S. Air Force, "Affordable Space Access: Issues, Choices, and Methods," *Space Workshop '95: Reducing Space Mission Costs*, Denver, CO, Mar. 8-10, 1995. Office of Technology Assessment, 1995.

stration which would prove the concept of single-stage-to-orbit."[73]

NASA has responded to this pair of directives by proposing what it believes to be a phased technology maturation program for the SSTO concept that periodically pauses along the way to evaluate its progress. If at any of the designated evaluation points the Administration decides that insufficient progress is being made, the pursuit of SSTO can be called off. NASA could then consider other RLV concepts and possibly draw from past SSTO technology development where applicable.

Some space policy analysts have taken issue with this approach. They argue that it grants—to the detriment of other RLV concepts—too much attention to the SSTO concept. They note that, because SSTO truly is a revolutionary goal, pursuing it with such vigor and then having to break off that pursuit in favor of upgrading the Space Shuttle or perhaps pursuing another RLV concept may lead to an inefficient and suboptimal result. For example, the addition of any type of first stage to a vehicle originally designed as an SSTO—whether it be a reusable booster, a set of strap-on solid rocket motors, or an expendable, liquid-fueled engine—would probably require a costly and time-consuming structural redesign in order to accommodate the additional thrust at takeoff.

One alternative to NASA's current RLV development strategy would be a more evolutionary approach—whereby "reusable building block" technologies would be gradually developed and tested in existing launch systems in the hope that they could eventually be expanded upon and used in an operational RLV (either SSTO or TSTO).[74] Another alternative would be to design, from the

[73] Ibid., sec. III(2)(b).

[74] Buzz Aldrin with John Kross, "Reusable Launch Vehicles: A Perspective," *Ad Astra*, March/April 1995, p. 31.

beginning, an SSTO capable of accommodating strap-on solid rocket motors to augment thrust. This has the advantage of avoiding the structural redesign noted above, and it would also provide a contingency for ensuring the nation's continued ability to design and produce long-range ballistic missiles.

Despite the criticisms, NASA believes that its current approach is prudent. It believes that SSTO is the RLV option that will reduce launch costs the most. Therefore, SSTO is deserving of a vigorous, yet carefully measured, technology development program—something NASA believes it has achieved with the X-33 program.

Criteria for Judging X-33 Program Success

Although the X-33 CAN sets out broad criteria for selecting a contractor for the fabrication and test flight phase (i.e., Phase II) of the program,[75] these criteria address the conditions for initiating Phase II and awarding a Phase II contract, but not for judging the success of Phase II.

The absence of specifically delineated criteria for evaluating the success of Phase II is troubling for two reasons. First, it may make it difficult for industry to predict how NASA will reach future X-33 program decisions. This uncertainty will affect the amount of money firms will be willing to invest in the X-33, if they choose to participate at all. Second, it will make it more difficult for NASA to explain to Congress and others why certain X-33 program decisions were made.

NASA, in conjunction with OSTP and the Office of Management and Budget (OMB), has established criteria to support both the 1996 and 2000 decisions.[76] They note that these criteria address cost, operations, and design and development factors and that all of the criteria (for both the 1996 and 2000 decisions) include a link to any follow-on commercial RLV development activities that might be proposed by prospective contractors.[77]

Some have suggested that because, in the future, Congress will be asked to provide continued annual support for the X-33 program, Congress may wish to ask that NASA provide a set of specific intermediate criteria for evaluating X-33 program success on an annual basis. Some analysts have noted, however, that annual evaluation of the X-33 program may slow the development process as a result of increased industry reporting requirements and the added dimension of program uncertainty that such evaluations would introduce.

If SSTO Commercialization Does Not Begin in 2000: What Next?

There is a distinct chance that industry will forgo commercial development of an RLV in 2000, either because the technical risks associated with the SSTO concept were not sufficiently resolved by the X-33 program or because market conditions do not justify the investment. If industry decides not to proceed with commercialization of an SSTO vehicle in 2000, one of five scenarios could unfold.

Scenario 1: **NASA decides to extend the SSTO development program beyond 2000 and upgrade the Space Shuttle to extend its operation until the newly projected date (sometime beyond 2012) for RLV replacement.** NASA might choose this course of action if it believes that the technical risks associated with SSTO are

[75] These criteria address cost, operations, and design and development factors, and all of the criteria include a link to any follow-on commercial RLV development activities that might be proposed by a prospective contractor.

[76] NASA, The White House, Office of Science and Technology Policy, and U.S. Office of Management and Budget, "Decision Criteria for the Reusable Launch Vehicle Technology Program Phases II and III," May 1, 1995.

[77] These criteria link the decision to proceed with Phase III of the X-33 program (i.e., development of a commercial follow-on RLV) to the ability of the X-33 and X-34 programs to meet "their respective program goals within a fixed Government budget [and demonstrate] that the industry-led, co-funded development of advanced space technology is an efficient, cost-saving program approach." Ibid., p. 12. Industry officials involved in the X-33 program are troubled by this link because they do not believe that X-34 program success is necessary to justify proceeding with Phase III of the X-33.

on the verge of being resolved. NASA's ability to choose this option would be constrained in two ways. First, NASA would have to convince Congress to fund both a continued SSTO development program and a Shuttle upgrade program that will probably cost more than current upgrade plans. Second, industry may decide that it is not in its long-term interest to pursue continued development of SSTO.[78]

Scenario 2: **NASA decides to initiate a partially or fully reusable TSTO development program starting in 2000 and upgrade to the Space Shuttle to extend its operation until the newly projected date (sometime beyond 2012) for RLV replacement.** NASA might pursue this option if it believes that the X-33 program achieved significant advances in RLV-related technologies, but decides that an SSTO vehicle remains beyond reach. NASA would probably request a more significant budget increase than that required for the previous scenario because NASA is skeptical that a TSTO designed to replace the Space Shuttle would be commercially viable and believes TSTO development would have to be fully funded by NASA. Without a significant budget increase, fully funding TSTO development might hamper NASA's ability to pursue the requisite Shuttle upgrades necessary to keep it flying safely until the TSTO comes on line.

The X-33 contractor would likely have an advantage over its competition in a TSTO competition if NASA could obtain from Congress the budget needed to fund the entire development of the vehicle. This advantage might be lessened if NASA, for whatever reason, turned to industry to partially or fully fund TSTO development.[79]

Were NASA to decide to select another contractor for the TSTO, it could resurrect NASA's dilemma over property and data rights for technologies developed during the X-33 program. NASA could exercise its title rights to inventions (and any associated data) developed in the X-33 program and transfer this knowledge to the new TSTO contractor. Alternatively, NASA could allow property and data rights issues to be worked out within industry. This, however, would raise costs and possibly slow down the development of the TSTO.

Regardless of the route that NASA takes to develop a TSTO under this scenario, some analysts believe that the resultant TSTO would be suboptimal because all prior development work will have focused on SSTO. Furthermore, they believe that the resultant TSTO would probably fall short of achieving the level of launch cost reductions that might otherwise have been achieved with a partially or fully reusable TSTO had the TSTO concept not been discounted by NASA earlier in the development process.

Scenario 3: **Industry decides to commercialize a partially or fully reusable TSTO that would allow NASA to retain its current plans for replacing the Space Shuttle in 2012 with the new RLV.** Industry might pursue this option if it thought that RLV-related technology advances achieved in the X-33 program were not enough to justify SSTO development, but did make TSTO a viable technology option. As noted in the discussion of the previous scenario, NASA officials question the commercial viability of TSTO. Therefore, industry would likely require a package of financial incentives—similar to, if not

[78] The X-33 contractor may decide that any continued investment in SSTO will not reap sufficient financial returns. Other potential contractors may conclude that the level of investment required for it to spool up an SSTO program and bring it to fruition by commercializing an SSTO vehicle would outweigh any potential financial reward from doing so.

[79] Any technical advantage gained by the X-33 contractor during development may be partially counterbalanced by the fact that the contractor invested substantial resources in a technology development program that did not result in an operational vehicle.

greater than, those it is currently demanding for SSTO development—before proceeding with commercialization of a TSTO vehicle.[80] Any TSTO developed under this scenario might be suboptimal for the same reasons articulated in the previous scenario.

Scenario 4: **NASA decides that it wants to initiate a program of block upgrades to the Space Shuttle that would extend its operations until 2020.** NASA would likely pursue this option if its current RLV technology development effort failed to reduce the technical risk associated with RLV launch systems sufficiently. NASA contends that wholesale block upgrades to the Space Shuttle would cost roughly $5 billion to $10 billion[81] on top of the current expenditures of about $4 billion per year for Space Shuttle operations. Improvements would likely include a handful of options listed in NASA's implementation plan as well as the incorporation of certain technologies successfully proven during the X-33 program (see section below on the Space Shuttle beyond 2000).

Scenario 5: **NASA decides to reevaluate its plans for the future, possibly commissioning a new space transportation study or perhaps considering alternative options already examined in past studies (e.g., development of a new, low-cost ELV that is capable of deploying either a cargo canister or a small, manned space vehicle).** NASA might pursue this option if it believes all other options are not viable. Continued Space Shuttle operation would depend on the operational status of the International Space Station. If operation of the International Space Station ceases, for any reason, Shuttle flights could potentially be suspended—which would enable NASA to direct more funds toward the development of alternative space transportation options. If Shuttle flights to the International Space Station are still necessary, the range of alternative space transportation options that NASA would be capable of pursuing may well be limited.

ROCKWELL INTERNATIONAL

Block upgrades initiated in 2000 could keep the Space Shuttle in operation until 2020.

ISSUE 1f: *Space Shuttle—beyond 2000*

If the government and industry decide in 2000 to forgo continued RLV development, NASA plans block upgrades to the Space Shuttle. These improvements are needed to ensure safe operations until 2020. At some point, however, the introduction of new technologies results in a substantially new vehicle with many of the same testing and safety concerns of a new vehicle. Furthermore, if reconstitution of old Space Shuttle production capabilities becomes necessary, it will require both time and money.

---

[80] Possible financial incentives include some type of guarantee from the government to launch its payloads exclusively on the newly developed TSTO. Another option would be to simply contribute government funds to the development of the vehicle.

[81] Daniel S. Goldin, NASA Administrator, testimony at a hearing before the Subcommittee on Space and Aeronautics, Committee on Science, U.S. House of Representatives, Washington, DC, Feb. 13, 1995. Some industry officials believe that this projection is an overestimate. One official suggests that Shuttle operations costs could be reduced by a $1 billion per year by privatizing the Shuttle and initiating a targeted upgrade program costing only $200 million per year.

NASA's implementation plan lists several potential upgrades that might be pursued. One of the main improvements proposed by NASA is the substitution of LFBBs for the existing solid rocket boosters (SRBs) between 2007 and 2010.[82] NASA touts the benefits of LFBBs in its implementation plan—namely increased safety, payload performance, and launch probability as well as reduced annual operating costs compared with SRBs. The implementation plan does not, however, outline any contingencies to address potential negative consequences associated with the switch over to LFBBs.[83]

Those concerned with maintaining a continued capability to produce ballistic missiles might have reason to be troubled by this omission. Replacing SRBs with LFBBs may have a significant impact on the solid rocket motor industry. NASA's implementation plan, however, does not address the national security implications of its SRB phase-out.[84]

In fact, NASA's implementation plan pays little direct attention to the Space Shuttle industrial base.[85] NASA's implementation plan seems to assume that the industrial base will possess the vitality to perform the block upgrade efforts cited by NASA as necessary to keep the Space Shuttle flying until 2020. OTA has spoken with a number of industrialists, however, particularly in the lower tiers, who are concerned with the present health of the industrial base and warn that reconstituting the Space Shuttle industrial base for the block upgrades requires more planning than NASA is currently doing.[86] The lack of planning suggests to some industrialists that NASA is not serious about upgrading the Space Shuttle beginning in 2000.

Finally, there remains the prospect of another Space Shuttle accident that results in the loss of an Orbiter. Past OTA analysis has shown that if Shuttle reliability is 98 percent there is a 50-50 chance of an Orbiter loss in the next 34 launches.[87] Such a loss would have major repercussions for both the Space Shuttle and X-33 programs.

▪ Fundamental Objective #2: U.S. Use of Foreign Launch Systems and Components

The NSTP expressly encourages federal departments and agencies "to take advantage of foreign components or technologies in upgrading U.S. space transportation systems or developing next generation space transportation systems."[88] The Administration's new policy also limits the flight of U.S. government payloads to U.S. space trans-

[82] NASA, op. cit., footnote 29, p. 21.

[83] NASA officials maintain that the decision to use LFBBs or make any other upgrade will not be made until the year 2000. They contend that at that time, as part of the decision process, a thorough analysis will be completed to assess the efficacy of changeover to LFBB or any other technology.

[84] See the discussion later in this report on preservation of long-range missile capabilities.

[85] NASA was unable to provide OTA with any studies by NASA of the Space Shuttle technology and industrial base. NASA, personal communication, March 1995.

[86] Recent discussion in the press may lend support to this concern. It seems that NASA might be considering consolidating Space Shuttle efforts by closing down Rockwell operations in both Downey and Palmdale, California and shifting all Shuttle related work to Kennedy Space Center in Florida. Although this is only one of several proposals currently under consideration by NASA, it suggests that major changes in the industrial base are inevitable. How these changes will affect NASA's ability to carry out block upgrades to the Space Shuttle appears to remain an open question. See United Press International, "NASA document targets Rockwell plants," Mar. 2, 1994

[87] At the time of OTA's analysis, experts considered Space Shuttle reliability to fall between 97 and 99 percent. See U.S. Congress, Office of Technology Assessment, *Access to Space: The Future of U.S. Space Transportation Systems*, OTA-ISC-415 (Washington, DC: U.S. Government Printing Office, April, 1990). NASA has made a variety of improvements to the Space Shuttle, some of which may improve its reliability and reduce the chance of losing an orbiter.

[88] The White House, Office of Science and Technology Policy, op. cit., footnote 3, sec. VI(2).

portation systems, in effect removing U.S. government payloads from the available international marketplace for launch services. In this, it follows past policy.

NSTP Fundamental Objective #2

Establishes policy on federal agencies' use of foreign launch systems and components. With the end of the Cold War, it is important for the U.S. to be in a position to capitalize on foreign technologies—including Russian technologies—without, at the same time, becoming dependent on them. The policy allows the use of foreign components, technologies and (under certain conditions) foreign launch services, consistent with U.S. national security, foreign policy and commercial space guidelines in the policy.

The NSTP recognizes the growing international interdependence of space activities by allowing launch of government payloads on foreign launch vehicles that are made available on "a no-exchange-of-funds basis to support the following: flight of scientific instruments on foreign spacecraft, international scientific programs, or other cooperative government-to-government programs."[89] This provision would cover, for example, such undertakings as the 1992 launch of the U.S.-French TOPEX-Poseidon spacecraft on an Ariane 4 launcher, and the use of non-U.S. launch vehicles during construction and operation of the International Space Station.[90]

The NSTP supports the negotiation of "international space launch trade agreements with other nations that define principles of free and fair trade for commercial space launch services."[91] It also notes that such agreements must conform with related U.S. obligations and treaties, such as technology transfer policies and the Missile Technology Control Regime (MTCR).[92]

ARIANESPACE

Ariane 4 in flight. The Ariane family of vehicles was primarily designed to serve the international commercial market.

The past two decades have brought striking changes in the character of space transportation services on the world market. The international marketplace for space transportation services has become far more complex and interdependent than it was just a few years ago. Before 1979, when the ESA successfully launched its first pay-

[89] The White House, Office of Science and Technology Policy, op. cit., footnote 3, sec. VI(1)(a).

[90] It also covers the planned launch of the Stratospheric Aeorsol and Gas Experiment (SAGE) and Total Ozone Mapping Spectrometer (TOMS) instruments on Russian Meteor 3 spacecraft, on Russian boosters.

[91] The White House, Office of Science and Technology Policy, op. cit., footnote 3, sec. V(1).

[92] The General Agreement on Tariffs and Trade (GATT) does not affect the international trade of launch services. Subsidy language in the Uruguay round of GATT pertains only to goods traded across borders—not services. Therefore, at present, the launch services market falls outside of the domain of GATT. Indeed, the unique nature of the launch service industry was a key factor behind the United States' pursuit and acquisition of launch service trade agreements outside of the GATT framework with both China and Russia.

load aboard the Ariane 1 ELV, the U.S. government was the only supplier of launch services for commercial satellites.[93] In the mid-1980s, the United States made the first moves toward commercializing its ELV fleet. In the late 1980s, first China, then the Soviet Union began to offer launch services aboard indigenous launchers. Most recently, as a direct result of the collapse of the Soviet Union, Russian and Ukrainian launch services and Russian subsystem technology have become available on the international marketplace.

These circumstances have forced policymakers to account not only for U.S. space transportation development needs, but also for the effects of the use of foreign technology in U.S. launch systems and competitive foreign launch services on the U.S. space transportation industry.

In particular, the high performance and low costs of Russian liquid-fueled engines and other launch technologies has led U.S. firms to consider incorporating Russian technology into current and future U.S. space transportation systems. Such uses of Russian technologies may reduce the costs and increase the performance of U.S. launch systems, making them more attractive to purchasers of U.S.-built satellites. Some policymakers and industry leaders, however, are concerned that incorporating Russian technology into U.S. launch systems might also lead to the loss of U.S. jobs in the space transportation industry. In addition, dependence on Russian technology raises concerns about the maintenance of the U.S. space transportation technology and industrial base, and U.S. readiness to meet national security challenges.

This section explores the effects that the policies toward introducing non-U.S. launch components into U.S. launch systems, contained in the NSTP and its supporting implementing plans, might have on the competitive position of the U.S. space transportation and satellite industries.[94] It also examines the use of non-U.S. launch services for U.S. commercial and foreign payloads and international trade in launch services.

Incorporating Foreign Technology into U.S. Launch Systems

Since the United States began work on the partially reusable Space Shuttle in the early 1970s, it has spent relatively little on the development of new technology for ELVs. Until the loss of *Challenger* in January 1986, the United States had followed a policy that focused on the Space Shuttle as the sole provider of launch services for all payloads.[95] The only new U.S. launch vehicle was the Titan IV, developed to provide an alternative means of placing large DOD payloads into orbit.

Although the manufacturers of ELVs have increased the payload capacity of U.S. systems and reduced operational costs, they have not kept pace with developments in foreign launch systems. U.S. launchers still take longer to process and to integrate with payloads than competing launch systems. The United States might be able to improve its fleet of ELVs and reduce launch services costs by adopting new technology and some of the automated launch operations practices of its competitors.[96]

The end of the Cold War has greatly broadened the available range of technology available to U.S.

---

[93] Both the United States and the Soviet Union also launched payloads for other governments, but arrangements were always made on a governmental level.

[94] See the section covering fundamental objective #4 for a discussion of the private sector role in space transportation decisionmaking and the status of the international market for launch services.

[95] This policy was first stated publicly in President Reagan's space policy of 1982. The White House, "National Space Policy Fact Sheet," Washington, DC, July 4, 1982.

[96] See, e.g., U.S. Congress, Office of Technology Assessment, *Reducing Launch Operations Costs: New Technologies and Practices*, OTA-TM-ISC-28 (Washington, DC: U.S. Government Printing Office, September 1988).

launch system manufacturers. During the Cold War, because of the close connection between launch vehicle technologies and ballistic missile technologies, the United States and other space-faring nations kept close control over the flow of launch technologies to other countries. As a result, U.S. insight into the details of the capabilities of the Soviet Union was quite limited.[97] U.S. and Soviet officials closely scrutinized even high-profile cooperative programs such as the Apollo-Soyuz linkup in 1975 to prevent unwanted technology transfers.

Beginning with the political reforms instituted during the Gorbachev regime, the United States began to gain direct access to Soviet technology, including launch technology. This process accelerated dramatically after the dissolution of the Soviet Union in December 1991. In the past few years, the United States has instituted closer relationships in space with the former Soviet Union (FSU), especially with Russia. These new cooperative relationships extend from large, complex partnerships like the International Space Station, to smaller, simpler cooperative agreements on space and earth science. Emblematic of the new relationships are the cooperative commercial agreements between U.S. and FSU firms, most of which involve launch technology, a field in which the Russians excel. U.S. firms have proposed employing Russian technology in U.S. boosters and have entered into agreements to market Russian boosters internationally.[98]

Incorporation of foreign technology might increase launch vehicle performance and reduce costs. The use of Russian technology promises to be particularly beneficial. The greatest strength of the Russian space program, and the principal strength of the Ukrainian program, lies in launch vehicles and associated technologies, particularly propulsion and rapid payload processing and integration. As noted above, several U.S. firms are exploring the use of Russian expertise and technology to enhance U.S. launch capabilities. Box 7 examines Russian launch technologies and outlines some U.S. efforts to incorporate them into U.S. space transportation systems.

U.S. Government Use of Foreign Launch Systems

As noted above, the NSTP explicitly allows the use of foreign launch systems on a no-exchange-of-funds basis to support cooperative programs. Such use can sharply reduce U.S. costs for scientific programs of interest to the United States.

The availability, robustness, and established reliability of Russian ELVs—built on large-volume, series production over many years—are also major assets for cooperative civil space activities. Their use on a no-exchange-of-funds basis could enable some projects that would not be undertaken otherwise.[99] The United States and Russia are pursuing this potential:

- Russian launch vehicles are being extensively scheduled to provide critical transportation for the assembly and operation of the International Space Station.
- The two governments are discussing the use of Russian launch vehicles in cooperative projects such as planned missions to Mars and Pluto.
- In 1991, a Russian Tsyklon booster (SL-14) and a Meteor-3 satellite carried a U.S. Total Ozone Mapping Spectrometer into orbit. By the end of the century, Russia will boost two additional Meteor-3 spacecraft carrying U.S. Earth observations instruments into space.

Such arrangements represent a way for Russia to make use of their substantial space capabilities

---

[97] Yet the relative openness of U.S. society always made U.S. technology developments more accessible to Soviet inquiry than the reverse.

[98] For a detailed discussion of these new cooperative relationships, see U.S. Congress, Office of Technology Assessment, *U.S.-Russian Cooperation in Space*, OTA-ISS-618 (Washington, DC: U.S. Government Printing Office, April 1995), chs. 3 and 6.

[99] This section does not address the potential risks of working with the Russians, especially economic and political instabilities, and the changing structure of Russian companies. See U.S. Congress, Office of Technology Assessment, op. cit., footnote 98, for such a discussion.

BOX 7: Russian Launch Technologies and U.S.-Russian Joint Ventures

Liquid-Fueled Engines

With the exception of the Space Shuttle Main Engine, the United States has developed no new liquid-fueled rocket engines since the 1960s.[1] Aerojet, a U.S. aerospace firm specializing in the design, testing, and manufacture of rocket engines, has proposed to buy a number of Russian NK-33 engines from the Samara Scientific and Technical Complex. These engines, which burn liquid oxygen and kerosene, were manufactured in the 1970s, but have features such as high chamber pressures that allow them to perform better than comparable U.S. engines. Aerojet believes that it can adapt the engines to make the U.S.-built Delta or Atlas boosters less costly and more powerful.[2] Similarly, Pratt and Whitney has proposed to modify the Russian RD-180 for use on the Atlas.[3] In each case the company involved expects that the cooperative venture will result in reducing the price of launch services.

U.S. industry and government officials have also explored the potential for incorporating modified Russian propulsion systems in U.S. X-vehicles. For example, Pratt and Whitney and NASA have announced plans to explore the application of tri-propellant rocket engine technology developed by NPO Energomash, to new RLVs.[4] Tri-propellant engines are capable of using both kerosene and liquid hydrogen oxidized by liquid oxygen. Such engines may offer the benefit of improving engine performance and reducing the size of propellant tanks, but also require greater logistics and operations complexity and additional propellant tanks.

Automated Launch Capabilities

The Russians perform their payload processing and integration and launch operations more efficiently and faster than U.S. launch providers. By doing much of their launch vehicle and payload integration off the launch pad, they are able to launch quickly and with less manpower than the United States, with no discernible loss of reliability (see table 12). Incorporating Russian operations methods and technology into U.S. launch operations could increase U.S. operability. Yet payload launch preparation and integration consume a large part of the time U.S. launch vehicles spend on the launch pad. To reduce pad time, U.S. operators would have to change the design of satellites and the methods used to prepare them for launch. U.S. firms would also have to redesign the launch vehicles themselves.

Advanced Materials

Russia also has extensive experience in the use of aluminum-lithium and other materials for space transportation applications, which are lighter in weight and more ductile than the conventional aluminum alloys used in the United States. This expertise allows manufacturers to use more cost-effective manufacturing processes, and to produce more durable engines.[5] Russian technicians have developed special

---

[1] The United States has, however, developed large solid rocket motors for the Space Shuttle, the Titan IV launch systems, and a variety of long-range ballistic missiles.

[2] Michael A. Dornheim, "Aerojet Imports Trud NK-33 Rocket Engine," *Aviation Week and Space Technology*, Oct. 25, 1993, p. 29.

[3] The RD-180 is derived from the larger, more powerful RD-170 that powers the first stage of the Russian-Ukraine Zenit booster. In November 1992, Pratt and Whitney signed an agreement with NPO Energomash to bring the latter's propulsion technology to the United States. See Jeffrey M. Lenorovitz, "Pratt Signs Accord with NPO Energomash," *Aviation Week and Space Technology*, Nov. 2, 1992, p. 25-26.

[4] Jeffrey M. Lenorovitz, "Tripropellent Engine Tested for SSTO Role," *Aviation Week and Space Technology*, July 11, 1994, p. 54.

[5] U.S. aircraft manufacturers have considered employing aluminum-lithium alloys in aircraft and NASA has funded a program to incorporate U.S. aluminum-lithium alloys in the Space Shuttle's external tank, in order to reduce the mass of the external tank. The lighter tank will allow NASA to place some 8,000 additional pounds of payload in the Space Shuttle when launched to the planned International Space Station.

BOX 7 (cont'd.): Russian Launch Technologies and U.S.-Russian Joint Ventures

coatings that permit engine turbine drive systems to run with a high proportion of oxidizer to fuel and increase engine thrust chamber durability. To date, the United States has not adopted the use of these materials because of the high cost of changing production lines. However, aluminum-lithium and other materials developed by Russian materials scientists might find future application in U.S. space transportation systems if they prove less costly than comparable U.S. materials.

Launch Technology Expertise

The Russian aerospace industry has many engineers and technicians with years of experience in areas such as rocket engine design, systems integration, and computer programming. The declining budgets for the aerospace enterprises mean fewer job opportunities and lower salaries for Russian aerospace engineers. Cooperative U.S.-Russia ventures could give the United States access to some of Russia's underutilized aerospace workforce, albeit at the expense of some American workers.

U.S.-Russian Joint Ventures

Rather than using Russian technologies in U.S. launch systems, some U.S. companies have sought to establish partnerships with Russian launch firms. Lockheed Corporation, for example, in January 1993 teamed with the Russian firms Khrunichev and RSC Energia in a joint venture (LKE International) to market launch services on the Proton launcher.[6]

In a similar arrangement, Boeing Commercial Space Development Company is seeking U.S. government approval for a joint venture with Ukraine's NPO Yuzhnoye,[7] RSC Energia, and Kvaerner A/S of Oslo, a Norwegian builder of offshore oil platforms, to market launch services using the Zenit vehicle, which is capable of placing 30,300 lbs of payload into LEO.[8]

---

[6] Jeffrey M. Lenorovitz, "Lockheed, Khrunichev To Market Proton Launcher," *Aviation Week and Space Technology*, Jan. 4, 1993.

[7] This proposal awaits licensing by the Department of State. See "U.S. Eyes Zenit Warily," *Space News*, Dec. 12, 1994, pp. 1, 28.

[8] The Zenit uses a highly automated launch processing system, which could give it a competitive advantage. This Russian innovation could, in principle, be applied to evolving U.S. systems as well.

SOURCE: Office of Technology Assessment, 1995.

and participate in cooperative space activities. In return Russian scientists would receive access to data that could enhance their capabilities in space science and applications. Potentially, the Ukrainian Zenit and Tsyklon boosters could be used for similar purposes, should Ukraine wish to participate in cooperative scientific activities.

Several U.S. instruments have already flown on European spacecraft, which were launched by Ariane launchers. In 1992, the U.S.-French TOPEX/Poseidon ocean topography spacecraft was launched into orbit aboard an Ariane 4 launcher. Europe has also sought to reduce its cash contributions to the International Space Station by offering

TABLE 12: Russian and U.S. Launch Processing Times Compared

Launch vehicle	Average total processing time (days)	Average time on launch pad (days)
U.S. Delta	70	40
U.S. Atlas	62	59
U.S. Titan IV	198	100
Russian Tsyklon	3	2 hrs
Russian Soyuz	18	2
Russian Proton	21	6
Russian-Ukraine Zenit	17	1

SOURCE: U.S. Department of Defense, 1994.

LOCKHEED MARTIN

Russia's Proton MLV is being marketed by LKE International, a U.S.-Russian joint venture.

the recently developed Ariane 5 launcher for carrying payloads to the space station.

Japan would like to employ its new H-II launcher to carry payloads to the space station. In early 1996, Japan plans to use its H-II to boost the Japanese Earth observation ADEOS spacecraft into orbit. ADEOS will carry two U.S. instruments.

International Trade in Launch Services

Beginning in the 1970s with the development of the Ariane launch system by ESA, the United States faced foreign competition in providing commercial launch services. Arianespace, S.A., the launch operations company incorporated in Europe, offers flights on the Ariane 4 series of MLVs and now commands about 60 percent of the world's available market for launch services.[100] In the late 1980s, Russia (then the Soviet Union) began to market the Proton, and China offered launch services on its Long March 2 and Long March 3. In 1992, LKE International began to market launch services on the Proton. In early 1995, Khrunichev and Daimler-Benz Aerospace formed Eurorokot, a consortium to market the Rokot SLV, developed from the SS-19 ballistic missile.

In the near future, firms in India, Israel, and Japan might begin offering launch services on the commercial market. The overall available market for launch services is currently small compared with the supply of launch vehicles (see table 13).[101] In the absence of new markets, the average available number of GEO launches in the early part of the next century will probably number about 15. The market is also highly cyclical in nature.

Each launch company competes for payloads from a different economic and political basis. Although the combined Russian and Chinese share of the world's market in launch services is currently quite small,[102] some U.S. launch providers fear

[100] The "available market" for launch services is composed of the entire market for which all U.S. and foreign commercial launch companies can compete. It includes U.S. and foreign commercial payloads and some non-U.S. government payloads, but excludes most government payloads of the launching entities.

[101] The market for payloads is currently a "buyer's market," in which the availability of launch systems exceeds the number of available payloads.

[102] Between January 1990 and September 1995, China captured about 10 percent of the commercial market for launch services. See U.S. Department of Transportation, Office of Space Transportation, "Quarterly Launch Report: April 1995," Washington, DC, April 1995. In the future, if existing launch agreements hold, China is likely to garner an estimated 15 percent of the commercial launch services market and Russia an estimated 10 percent.

TABLE 13: International Launch Firms Offering Commercial Launch Services

Country	Firm	Launch vehicle
China	China Great Wall Industry	Long March 2 Long March 3
Europe	Arianespace	Ariane 4 Ariane 5
Russia	LKE[a] Eurorokot[b] STC	Proton Rokot Start-1
United States	McDonnell Douglas	Delta
	Lockheed Martin	Atlas LLV
	Orbital Sciences	Pegasus Taurus
	EER Systems	Conestoga

[a] U.S.-Russian joint venture offering Russian vehicle.
[b] Russian-German joint venture offering Russian vehicle.
SOURCE: Office of Technology Assessment, 1995.

that low prices made possible by a combination of extremely low wages and a non-market economy will undercut U.S. opportunities to sell launch services to the available market. U.S. satellite manufacturers, however, generally support the entry of Chinese and Russian launchers into the world market because these launch systems provide a wider choice of launch schedule and performance, reduced launch prices, and incentive for U.S. launch providers to lower their prices.[103]

Although all launch providers receive some government support, the arrangements differ among countries. These differences, and disagreements over what effects such differences make in an international competition for launch services, make it extremely difficult to achieve genuinely free and fair trade in the commercial launch services market (see box 8).

The perceived degree of European subsidy for Ariane and the effects of non-market competition from China and Russia have led to claims of unfair competition by U.S. launch service firms. Other countries respond that not only has the United States subsidized its launch vehicle development, it also maintains a relatively large, protected market of government launches for DOD and other government departments and agencies. However, other governments protect part or all of their launch services market as well. ESA satellites generally fly only on the Ariane launcher, and Chinese and Russian government payloads fly only on indigenous launchers. All these concerns remain points of contention as China, Europe, Russia, and the United States attempt to reach agreement over appropriate mechanisms to manage international trade in launch services.

The Missile Technology Control Regime

Managing trade in launch services is complicated even further by the added dimensions of arms control, nonproliferation, and the control of technology transfer. The NSTP acknowledges this added dimension by requiring that:

> International space launch agreements in which the U.S. is a party must be in conformity with U.S. obligations under arms control agreements, U.S. nonproliferation policies, U.S. technology transfer policies, and U.S. policies regarding observance of the Guidelines and Annex of the Missile Technology Control Regime (MTCR).[104]

Although missile systems and space-launch systems serve quite different purposes, they have much in common. Hence, the United States has a strong interest in limiting the ability of countries that currently do not possess missile capabilities from acquiring space launch technology. U.S. officials worry, in particular, that the testing and development of weapon delivery systems can be

[103] Warren Ferster, "China Wins Big in Launch Deal," *Space News*, Feb. 6, 1995, p. 1, 20.

[104] The White House, Office of Science and Technology Policy, op. cit., footnote 3, sec. V (1) (b).

BOX 8: Government Involvement in Space Transportation Development

Europe

The European Space Agency paid for the full development costs of the Ariane family of launch vehicles and the associated infrastructure. From the beginning, ESA planned to operate the Ariane system as a commercial launcher,[1] and set up Arianespace, S.A., a European firm incorporated in France, to market and operate launch services.[2] Through thoughtful, cost-effective design and aggressive marketing, Arianespace is now the industry leader in providing commercial MLV services.[3] ESA continues to fund improvements to the Ariane system; most recently it has paid for the development of the Ariane 5 launch vehicle, launch pad, and infrastructure in Kourou, French Guyana.

United States

The federal government developed the Atlas, Delta, and Titan launch systems in the 1960s and continued to improve them through the 1970s.[4] The United States also began to develop the Titan IV HLV in the mid-1980s. In 1983, the Reagan Administration made the Atlas, Delta, and Titan III launch systems available for private ownership and operation,[5] but the development of private launch services was inhibited by competition with the government-operated Space Shuttle.

Private sector entry into the commercial market for launch services became economically feasible only after the Reagan Administration limited use of the Space Shuttle to payloads that require the unique features of the Shuttle.[6] Several U.S. firms now offer commercial launch services (see table 13).

Both NASA and DOD continue to fund limited launch system improvements that will benefit the U.S. government by improving performance or reducing costs. NASA contributed approximately $54.9 million to such development work in FY 1994 and plans to spend $33.6 million in FY 1995. The Air Force spent about $191.5 on ELV technology development in FY 1994 and will spend about $149.2 in FY 1995.[7]

---

[1] ESA followed this route in order to 1) establish Europe as a supplier of commercial space services, and 2) help keep launch costs down.

[2] Arianespace's principal investors are 35 European companies, 13 European banks, and CNES, the French Space Agency.

[3] Arianespace received a boost in sales following the loss of *Challenger* when the United States decided not to offer commercial launch services on the Space Shuttle. Before then, the U.S. government had been competing with both the European Ariane and U.S. commercial launch service providers.

[4] Each of these launch vehicles derives from ballistic missiles.

[5] On May 16, 1983, the Reagan Administration announced that "the U.S. Government fully endorses and will facilitate commercial operations of [ELVs] by the U.S. private sector."

[6] Ronald Reagan, Presidential Decision, Aug. 15, 1986.

[7] These budget figures include $50.7 million in FY 1994 and $67.0 million in FY 1995 for the Range Standardization and Automation program. Personal communication, U.S. Department of Defense, May 1995.

achieved under the guise of developing a space launch program. Any country that can design and build space launch vehicles should be considered capable of developing ballistic missiles.

In 1987, in order to limit proliferation of long-range delivery systems capable of delivering weapons of mass destruction (nuclear, chemical, and biological weapons), the United States and other Western industrialized nations developed the MTCR. Admittance to the U.S. satellite market has become a tool in encouraging adherence by China and Russia to the MTCR. Recently, the members of the MTCR have encouraged Russia and Ukraine to join the MTCR. Russia, for example, has agreed to abide by MTCR rules until it becomes a full-fledged member. Participation in the

BOX 8 (cont'd.): Government Involvement in Space Transportation Development

Russia

The Soviet Union developed the world's first space launch system in the late 1950s, derived from its ballistic missile systems. Since then, the Soviet Union, and now Russia, has led the world in the number of launches per year. Until very recently, the Soviet (now Russian) government funded all launch system development, manufacturing, and operations. Now, as a result of the collapse of the Soviet Union and the moves by Russia to develop a market economy, Russia has privatized much of its space industry, and markets a variety of launch services to the international community.[8]

On the international market, Russian firms benefit from low labor costs and large-scale, assembly-line production. Although major elements of Russia's command economy still exist, the new space enterprises have begun to assume the costs of manufacturing. Whether the firms or the Russian government will shoulder the burden of new developments required to meet market needs for launch services is not yet clear. Nevertheless, financial analysts experience great difficulties gaining insight into development and production costs, in part because Russian officials themselves often do not know how much goods and services cost. Hence, it is extremely difficult to know how Russian pricing practices relate to the cost of providing services.

China

The Chinese introduced their Long March series of launchers to the world market in 1985.[9] Like the ELVs of other countries, China's launchers were developed from ballistic missiles.[10] Marketed through the government company, China Great Wall Industry Corporation, the Long March 2 and Long March 3 have been used primarily to place communication satellites into GEO. Since China began to offer services to the international community, it has successfully launched five satellites into space.[11] Although China Great Wall Industry operates much like a private corporation, the costs of launcher development and manufacturing are borne by the Chinese government. China has the competitive cost advantage of very low wages and non-market accounting practices.

---

[8] Russia launched its first U.S. satellite on January 23, 1995 aboard a Kosmos launcher. See James R. Asker, "Russia Launches Its First U.S. Satellite," *Aviation Week and Space Technology*, Jan. 30, 1995, pp. 68-69.

[9] "China Offering Space Launch Services to International Users," *Aviation Week and Space Technology*, Apr. 8, 1985, p. 25.

[10] China received the basic ballistic missile technology from the Soviet Union in the 1960s.

[11] Long March vehicles have also sustained several failures; the most recent was a failure of a Long March 2E on Jan. 26, 1995, which was carrying an American Hughes satellite, Apstar 2.

SOURCE: Office of Technology Assessment, 1995.

MTCR requires that Russia prohibit the transfer of complete missile systems, components that could be used to make complete systems, and technology involved in the production of components or of complete systems. Ukraine has agreed to abide by the terms of the MTCR to demonstrate that it would follow the MTCR upon admittance to the regime.

China has refused to join the MTCR, although it too has agreed to abide by its terms. The United States monitors China's compliance to the MTCR and has raised issues of noncompliance. On October 4, 1994, the United States and China agreed to "work together to promote missile nonproliferation through a step-by-step approach to resolve differences over missile exports."[105] The United States and China agreed to "hold in-depth discussions on the MTCR. . .[and] to promote eventual Chinese membership in the MTCR."[106] Under the terms of the MTCR and U.S. law governing sanctions against foreign entities, the United States could levy sanctions against a Chinese launch company, including prohibition of satellite launches, if the United States found that the entity was selling missile-related technology to a country that did not previously possess such technology.

Although the MTCR has had some measure of success in limiting the flow of missile-related technology, it also inhibits the flow of technology that could be used to develop new launch vehicles. Officials in India and Brazil, for example, have complained that their efforts to develop indigenous launch vehicles have been inhibited by the MTCR. Recently, the Brazilian government has agreed to join the MTCR, in part because member countries had limited technology transfer to Brazil. This move may help Brazil obtain crucial guidance technology for development of its indigenous launch vehicle.[107]

Space Launch Trade Agreements

The United States has considerable influence over trade in launch services because it continues to sell more satellites on the international market than any other country, even as its share of the launch services market has declined. When a U.S. satellite firm offers to sell a payload to a foreign company or government entity, it specifies launch services as part of the package. These services are selected in the international marketplace based on several factors, including estimated reliability, success in meeting schedules, and price. Typically, U.S. satellite firms may sell a satellite to a foreign firm or government and launch it on a third-party launch vehicle.[108]

Because the United States could otherwise severely restrict the international sale of U.S.-manufactured satellites launched on other countries' vehicles, the United States has been able to negotiate space launch trade agreements with both Russia and China. The fundamental premise of these trade agreements is "to establish criteria regarding participation by space launch industries in countries in transition from a non-market to a market economy."[109] The office of the U.S. Trade Representative (USTR)[110] is the U.S. agent in these negotiations.

The United States has depended on quantity restrictions and pricing guidelines to manage the impact of Chinese and Russian launch services on U.S. space transportation services providers. The following paragraphs summarize the contents of the existing and past trade agreements.

The 1989 U.S.-China Launch Agreement: The United States reached its first launch services agreement with China in January 1989. It remained in force until December 31, 1994, and established the fundamental structure for later agreements with both China and Russia. The agreement allowed only nine Chinese launches of international payloads to GEO over the period of

[105] U.S. Department of State, Office of the Spokesman, Fact Sheet, "Joint United States-People's Republic of China Statement on Missile Proliferation," Washington, DC, Oct. 4, 1994.

[106] Ibid.

[107] Philip Finnegan, "Brazil Prepares to Sign MTCR," *Space News*, Apr. 24, 1995, p. 3, 29.

[108] For example, the Chinese have scheduled AsiaSat 2, a satellite built by Martin Marietta (now Lockheed Martin) for launch later this year on a Long March 2E launcher. If successfully launched, AsiaSat 2 will be owned by Asia Satellite Telecommunications Co. of Hong Kong.

[109] The White House, Office of Science and Technology Policy, op. cit., footnote 3, sec. V(1).

[110] The various implementation plans submitted by DOD, NASA, and DOT/DOC do not speak to the issues of free and fair trade, since those negotiations are the domain of the USTR. The DOT/DOC implementation plan does cite, however, the importance of trade agreements to limit market perturbations.

the agreement.[111] Additionally, the agreement essentially precluded Chinese launch service providers from collectively launching more than three international payloads in any one calendar year.

According to the agreement, both the United States and China support "the application of market principles to international competition among providers of commercial launch services, including the avoidance of below-cost pricing, government inducements, and unfair trade practices."[112] These premises were the basis for a pricing guideline known as pricing "on a par." In effect, the guideline required that Chinese launch service providers establish prices, terms, and conditions in a manner comparable to the prevailing norms of the international launch services market.[113] Failure to price on a par, according to the agreement, would be grounds for punitive actions by the United States as permitted by U.S. laws and regulations. The agreement did provide for the less stringent option of consultation on demand by either party. However, it failed to establish criteria by which China could be judged to be in or out of compliance with the provision of pricing on a par. Therefore, the provision had little effect on Chinese behavior in the market, although the overall agreement may have limited the number of commercial satellites launched by the Chinese.

GREAT WALL INDUSTRY CORPORATION

The Long March 2E. China began marketing its Long March family of launch vehicles in the late 1980s.

The 1993 U.S.-Russia Launch Agreement: In September 1993, the United States and Russia signed a launch agreement that is similar to the U.S.-China agreement, but which adds an additional measure of control. The agreement, which lasts until December 31, 2000, establishes pricing criteria on which the actions of Russian launch service providers can be judged on a bid-by-bid basis. Any Russian bid more than 7.5 percent below the lowest Western bid for a comparable launch service triggers automatic consultations between the two parties.[114] When such a consultation is called, Russia is given the opportunity to provide valid

[111] The agreement provided for annual consultations at which the quantity restriction could, at the request of China, be adjusted because of changed conditions in the market for launch services.

[112] "Memorandum of Agreement Between the Government of the United States of America and the Government of the People's Republic of China Regarding International Trade in Commercial Launch Services," Washington, DC, Jan. 26, 1989, pp. 1-2.

[113] Ibid., p. 2.

[114] "Agreement Between the Government of the United States of America and the Government of the Russian Federation Regarding International Trade in Commercial Space Launch Services," Washington, DC, September 1993, art. V(2).

reasons (e.g., different insurance conditions or additional integration costs) for pricing below the 7.5 percent differential.

The Russian agreement allows for the launch of eight principal payloads to GEO or geosynchronous transfer orbit (GTO) over seven years[115] and limits Russian launch service providers, as a group, to two launches within any 12-month period. This provision is notably more restrictive than the equivalent one in the 1989 U.S.-China agreement, which limited China to no more than three launches in any one calendar year.

U.S. and Russian officials disagree over which launches count against the agreement's quota—a disagreement that stems from a practice known as "leasing on orbit," whereby a satellite owned by a firm from one country and launched by that country is leased to a foreign entity after the satellite reaches orbit. Russian officials contend that leasing Russian satellites launched by Russian launchers does not count as the launch of an international payload and thus should not be counted against the quota. U.S. trade officials strongly disagree, insisting that such launch opportunities should count against the quota on grounds that they should be open to international competition.[116] Trade officials have also not reached agreement on how to count the launch of multiple satellites on a single launcher.

Unlike the 1989 U.S.-China agreement, the agreement between the United States and Russia requires the two countries to consult about commercial launches to orbits other than GEO and GTO, and suborbital launches on a case-by-case basis, including the emerging market of launches to LEO.[117] However, the agreement establishes no specific quantity restrictions or pricing guidelines for LEO launch services.

The 1995 U.S.-China Launch Agreement: In March 1995, the United States and China entered into a new launch agreement that will last until December 31, 2001. In developing this agreement, U.S. officials benefited from experience with the earlier launch agreements and addressed three U.S. concerns: the appropriate level of quantity restrictions and pricing guidelines, leasing on orbit, and the emerging LEO market. U.S. satellite firms and U.S. partners of Russian launch services firms had criticized the U.S.-Russia agreement for its quantity restrictions and pricing guidelines,[118] arguing that the quantity restrictions deny U.S. satellite manufacturers adequate access to highly reliable Russian launch services and that the 7.5 percent consultation trigger increases Russian launch service prices.

In response to these criticisms, and recognizing that the new Chinese agreement would probably set precedents for revising the U.S.-Russia agreement, the USTR added to the new agreement a built-in adjustment to the quota in the event that the GEO payload market improves. This provision allows China a total of two additional GEO launches if, over the first three years of the agreement, an average of 20 or more GEO payloads per year are launched in the available world market. It allows for an additional three launches (for a total of five extra launches), if this trend continues for a fourth year.

The USTR relaxed the previous 7.5 percent level for consultation to 15 percent. It was able to do this because the new Chinese agreement also delineates a set of quantified "comparability factors" to which both the United States and China can refer in order to determine whether a Chinese bid below the 15 percent level is justified. This

115 Excluding the INMARSAT 3 and three Iridium LEO satellite launches that Russia had contracted with consultations before signing the agreement, four of the eight GEO/GTO launches may consist of two principal payloads on a single launch vehicle.

116 The United States did not count, however, the first such launch.

117 The agreement specifically mentions the Iridium communications constellation currently under development by Motorola and its partners.

118 Brian D. Dailey, "U.S. Trade Policy's Future Role," *Space News*, Oct. 17, 1994, p. 15.

method is intended to account for differences in business practices and market environments (e.g., payment terms and conditions, insurance costs, intended orbit, or major differences in launch procedures) that would change the effective price to the customer.

The new Chinese agreement clearly spells out that a satellite leased on orbit counts against the quota. Although the USTR believes that China is unlikely to lease communication satellites on orbit because the country currently lacks adequate communications satellite capacity, it sought this clarification in order to provide leverage in later negotiations with either Russia or China.

Finally, the latest U.S. agreement with China states that China's participation in the LEO market should be proportionate, nondisruptive, and not interfere with U.S. participation in the LEO market. The agreement requires consultations if China contracts to launch over 50 percent of any given LEO communications constellation.

Future Possible Agreements: U.S. officials expect Russian trade officials to request changes in the present launch agreement with the United States. In particular, Russia is likely to push to have the LEO provisions of the agreement brought into line with those delineated in the latest U.S.-China agreement. Russia, with the strong support of U.S. satellite manufacturers and U.S.-Russian launch company partners,[119] may also seek an increase in its quota[120] and a relaxation of the 7.5 percent consultation level.

Recently, the United States invited Ukrainian officials to begin exploratory negotiations toward establishing a bilateral launch services agreement similar to those negotiated with China and Russia.[121] Such an agreement could allow U.S. satellite manufacturers the choice of the Ukrainian Zenit and Tsyklon vehicles for placing payloads into orbit. Reaching agreement with Ukraine would place even more pressure on the United States to modify its agreement with Russia.

Issues for Congress

The use of foreign launch technology and foreign launch services and cooperation between the United States and other countries raise several issues of interest to Congress. These include the effects on U.S. industry, the U.S. technology base, technology transfer, and possible rules for international trade in launch services.

ISSUE 2a: *Effects of the use of foreign launch technologies on the U.S. aerospace industrial base*

Provisions of the NSTP regarding U.S. use of foreign technologies in U.S. space transportation systems are designed to improve the efficiencies of U.S. launch systems, in order to meet the domestic need for access to space and compete more effectively in the international space transportation market. As noted earlier, Russian propulsion technologies are of greatest interest to U.S. firms. Because the requirements of the Soviet/Russian space program have differed from those of the U.S. program, Russia has developed systems with different operational and design characteristics.

Access to Russian technological innovations could offer U.S. manufacturers a wider range of design possibilities from which to choose, many of which have already been tested and implemented by the Russians. U.S. officials have also expressed interest in adopting some of the technologies and techniques used by Arianespace, Russia, and Ukraine for launch operations. Recently, an Arianespace official offered the Ariane 5 heavy-lift launcher as a candidate for the DOD

---

[119] "We still believe the best outcome [in launch services agreements] is the removal of quotas in such agreements to best serve the [U.S.] satellite community since they constitute 90 percent of the space market," Brian Dailey, Lockheed vice president for Washington operations, quoted in Warren Ferster, "China Wins Big in Launch Deal," *Space News*, Feb. 6, 1995, pp. 1, 20.

[120] Peter B. de Selding, "Russian Quota Questioned," *Space News*, Apr. 3, pp. 4, 45.

[121] Warren Ferster and Peter B. de Selding, "Zenit, Cyclone Parley To Start," *Space News*, Apr. 10, 1995, pp. 1, 20.

AEROJET

The Russian NK-33 engine is one candidate for incorporation into U.S. space transporation systems.

EELV program, arguing that the United States could thereby avoid a costly development program.[122]

The effect on the U.S. aerospace industry of using Russian technology in U.S. launch vehicle systems will depend on how U.S. firms structure commercial cooperation with the Russians and on which part of the industry attention is focused. On the one hand, access to different and up-to-date technologies, production and processing methods, and cheaper hardware could make the U.S. aerospace industry stronger in an increasingly competitive world market for space-related services. On the other hand, cooperative arrangements could also lead to unwanted technology transfer, strengthening of a competitor, loss of domestic production jobs, and a weakening of U.S. capabilities because of dependence on a foreign source.

Representatives of some lower-tier firms in the U.S. launch industry expressed another viewpoint at a recent OTA workshop. They believe that the domestic launch industry is struggling and does not need another competitor in the medium-to-heavy launch service market, irrespective of any possible enhancement of U.S. capabilities through cooperation with the Russians.[123]

U.S. firms could adopt several approaches to using foreign technology. One approach is to buy components directly from foreign suppliers. In the case of Russian propulsion systems, for example, such a buy would probably result in job losses for the engine-manufacturing segment of the domestic industry. Yet, in most cases, testing and systems engineering will still be required. Also, cheaper engines might make U.S. launch services more competitive, potentially increasing business and creating jobs in that sector of the industry, as well as in others stimulated by low-cost launch services.

Alternatively, a U.S. firm could buy a license for a given engine technology and set up its own production line. Licensing of technology would result in increased employment for U.S. workers if the licensing firm is successful in producing a product. It could also make those parts of the industry that depend on the licensed technology more competitive in the world market.

Ultimately, as the worldwide launch industry becomes more like other commercial industries, the use of foreign components and systems will become more widespread, although the terms of the MTCR will limit technology transfer. If the U.S. launch industry is to become competitive on the world market, it may have to become more flexible and make effective use of non-U.S. technologies. Currently, the United States leads in the development of avionics, computers, electromechanical actuators, and other technologies that support the launch industry. U.S. industry could also improve its competitive stance by developing launch technologies for non-U.S. launch systems. Exporting many of these technologies will require the relaxation of U.S. export controls.

---

[122] Ben Ionatta and Cheri Privor, "Arianespace's EELV Proposal Finds Little Favor," *Space News*, Apr. 10, 1995, p. 3. The official listed several options, among which is the potential for licensing Ariane 5 technology to U.S. firms for construction in the United States.

[123] A background paper on this workshop is forthcoming, June 1995.

The United States must also decide how much of its industrial base should be maintained to ensure access to space and to meet national security needs. Use of foreign technologies could reduce the amount of R&D required of U.S. firms,[124] resulting in reduced costs, but it could also undercut the development of U.S. capabilities in certain areas. Because the space industry is considered to be indispensable to the security of the United States, many argue that the United States should develop and maintain its capabilities in certain critical areas to prevent any weakening in its technological base.

DOD is willing to use launch systems that have foreign components and technology, but only in such a way that foreign suppliers cannot deny DOD access to space:

> This can be accomplished by such measures as stockpiling critical foreign components and assuring that alternative sources of critical components could be developed in a timely fashion should foreign sources cease to be available.[125]

Although this approach might result in higher costs to the government, it ensures that the United States will be able to fulfill its space-related national security needs without depending on foreign suppliers of launch services. Arianespace has suggested that if the United States wished to pursue the use of Ariane 5 (developed by U.S. allies) for the EELV program, it could license the entire launcher for construction in the United States, modifying it as needed and buying some parts in Europe. Such an arrangement could substantially reduce the costs of building and operating a U.S. heavy-lift launch vehicle.[126] However, building a vehicle under license might inhibit the development of new U.S. technology that could be used to improve the U.S. MLV fleet.

ISSUE 2b: *The ground rules for international trade in launch services*

The United States' response to the competitive challenge posed by the market entry of Chinese and Russian launch systems has been to work with trade representatives of other countries to achieve a common understanding of what constitutes fair trade practices in selling launch services:

> The U.S. and the PRC [China] support the application of market principles to international competition among providers of commercial launch services, including the avoidance of below-cost pricing, government inducements, and unfair business practices.[127]

The resulting launch service agreements limit the total number of commercial launches China and Russia can sell on the world market. In allowing each country to sell launches to U.S. satellite firms, but limiting them to a fixed total of the world's available launch services market over a defined period, the USTR has steered a middle course between a "hands-off" approach and allowing no foreign launches of U.S.-built satellites. In other words, the USTR has attempted to manage the commercial market in launch services.

Critics of this policy, who include U.S. satellite manufacturers and satellite customers as well as U.S. partners of Russian launch firms, complain that such market management effectively raises the overall price of launch services. They argue that because the agreements limit the total number of Chinese and Russian launches between now and 2001,[128] U.S. launch providers have little in-

[124] U.S. industry could use the technologies to develop the base for the next generation of R&D. However, it might have less incentive to invest in R&D if it were able to earn sufficient profit with foreign technology.

[125] U.S. Department of Defense, "DOD Implementation Plan for National Space Transportation Policy," PDD/NSTC-4, Washington, DC, Nov. 4, 1994, p. 9.

[126] Ben Ionatta and Cheri Privor, op. cit., footnote 122.

[127] Op. cit., footnote 112, art. II(A).

[128] The U.S.-Russia agreement concludes on December 31, 2000; the U.S.-China agreement concludes on December 31, 2001.

centive to reduce their prices. On the other hand, U.S. launch services companies worry that without some limits on China's and Russia's ability to sell launches below cost, U.S. market share will slip even further than it has.

The USTR has attempted to structure the launch service agreements to encourage China's and Russia's moves away from their bureaucratic command systems toward adopting the financial practices of market economies, in which prices are based primarily on actual costs of labor, manufacturing, and operations, while encouraging a viable U.S. space transportation industry, as called for in the NSTP.[129]

Some observers worry that the present launch agreements with Russia and China do little to promote a change from centrally managed economies. They argue that the existing framework, based on quantity restrictions combined with bid-by-bid pricing guidelines, lacks the ability to foster market-oriented behavior. These skeptics contend that the existing agreements simply hand a portion of the launch services market to the Russians and Chinese in exchange for their agreement not to undercut U.S. launch service providers with very low launch prices, subsidized by other parts of their economies.[130] As they see it, the best long-term protection available to U.S. launch service providers would be a marketplace where all players engage in fair competition based on market principles—something they do not believe the present agreements are designed to achieve. Competing on the basis of market mechanisms has the benefit that a firm cannot long continue to offer services substantially below cost and survive.

One of the major obstacles in policing the launch services agreements is the difficulty of determining the actual price of a launch. Potential customers must take many factors into account, including demonstrated launch success rate, the condition of payload integration facilities, geographical location of the launch pad, and cost of insurance. For example, a $50-million Atlas launch may provide more or less value than a $50-million Ariane launch, and price alone will not determine the winning bid.

Despite these criticisms, some argue that the current launch agreements with Russia and China are the best that can be expected under the circumstances. According to this view, trade agreements of this sort are often cumbersome and disagreements over details are inevitable. Moreover, they often include foreign policy considerations beyond the narrow scope of the agreements.

Competition between Arianespace and U.S. launch firms presents the USTR and U.S. launch firms with a set of issues different from those raised in agreements with China and Russia. Although European governments generally maintain much closer relationships with their major industrial firms than does the U.S. government, European market norms and practices are much closer to the U.S. example than are those of China and Russia.

Nevertheless, U.S. launch companies complain that the European governments unfairly subsidize Arianespace's operation.[131] Attempts to determine the amount and nature of these subsidies have generally concluded that, although such subsidies may be greater than U.S. subsidies, determining the exact amount and how they may affect the pricing of launch services is extremely difficult. The structural differences between U.S. and European institutions, coupled with the financial complexities of a launch services agreement, cloud the comparison of subsidies. Hence, suggestions to impose sanctions on European launch services have been short lived. In 1984, for example, Transpace Carriers, Inc., filed a com-

---

[129] The White House, Office of Science and Technology Policy, op. cit., footnote 3, sec. IV(1).

[130] Andrew Lawler, "Industry Criticizes U.S. Launch Agreements," *Space News*, Oct. 3, 1994, p. 3.

[131] See "U.S. Space Launch Services Company Brings Unprecedented Complaint Against Europeans," *U.S. Import Weekly* 9:1088, 1984.

plaint under section 301 of the U.S. Trade Act of 1974, stating, among other things, that Arianespace was unfairly subsidized compared with U.S. carriers.[132] Investigation following the complaint found that European subsidies and pricing practices were not out of line in comparison to U.S. practices.

In addition, U.S. satellite manufacturers tend to oppose any type of pricing agreement with Europe, seeing such an agreement as a barrier to globally provided launch services, and an opportunity for both Europe and the United States to raise prices artificially. Ultimately, the United States may find it more effective to concentrate on ways of reducing the costs of U.S. space transportation systems, rather than attempt to manage the international market in launch services.

ISSUE 2c: *Controlling technology transfer and other foreign policy objectives*

Cooperative ventures entail the risk of transfer of domestic technologies that could be used to strengthen a competitor's position in the international aerospace market and could enable the development of ballistic missiles. Experts disagree over how effective means to prevent such transfer can really be, but present policy clearly moves toward loosening trade restrictions. Specifically, many components having to do with satellites and satellite technology have been moved from the U.S. Munitions List[133] onto the Commerce Control List, effectively making it easier to trade in those items.

Recent reports suggest that the State and Commerce Departments are working on loosening restrictions further.[134] As noted above, the greater relaxation of these restrictions could result in improved U.S. trade in launch technologies. On the other hand, the United States must remain sensitive to the potential proliferation of technologies that would assist belligerent countries in developing the means of delivering weapons of mass destruction.[135]

The desperate economic state of Russia and Ukraine makes the sale of expensive, high-technology missile components and systems to other countries extremely attractive. For example, in 1992, India contracted with Russia to buy a liquid-oxygen/liquid-hydrogen-fueled engine to be used as the upper stage for its Geosynchronous Satellite Launch Vehicle. The United States opposed the sale on grounds that it violated the MTCR, a move that both India and Russia resisted. U.S. officials were concerned that the technology associated with the engines would assist India in building ballistic missiles. Fearing that the United States would institute sanctions allowed by the terms of the MTCR, in 1993 Russia agreed to break its contract with India and withhold the engine technology.[136]

Even if Russia abides by the MTCR and prohibits the export of hardware useful in ballistic missiles, it might not be able to prevent the emigration of rocket scientists to countries seeking to use their expertise. Despite Russia's apparent concern over the loss of its aerospace engineers, it might not be able to prevent the departure of many to countries that might be hostile to U.S. interests. People with expertise can freely emigrate from Russia to the neighboring Newly Independent States, and keeping track of where they go from there might not be possible.

The Clinton Administration considers that assisting the Russian civilian space program to stay

---

[132] Ibid.

[133] 22 CFR Ch. 1, Subchapter M-International Traffic in Arms Regulations, Part 121-The United States Munitions List

[134] "Satellite Export Controls to Ease," *Space News*, Feb. 20, 1995, p.1

[135] U.S. Congress, Office of Technology Assessment, *Export Controls and Nonproliferation Policy*, OTA-ISS-596 (Washington, DC: U.S. Government Printing Office, May 1994).

[136] Experts differ in their opinions about the usefulness of cryogenic engines for weapon systems. Weapon systems benefit from constant readiness, and cryogenic engines take a long time to prepare for launch. Still, all early U.S. ballistic missiles were liquid-fueled.

as healthy as possible and capable of retaining its experts will be in the interest of global nonproliferation. Similarly, the United States provides some direct funding to scientific researchers responsible for the development and engineering of nuclear, chemical, and biological weapons in an effort to keep them employed in areas other than the development of those weapons.[137]

Many of the scientists and engineers in the Russian civil and military space programs have expertise that could be usefully applied to space science missions. The incorporation of Russia in the International Space Station has been undertaken in part to support Russia's attempts to maintain its civilian space efforts.[138] Even during Cold War periods when the political atmosphere made larger, high-profile cooperative science efforts unacceptable, small, low-profile science projects involving Russian and U.S. scientists continued. That ongoing cooperation kept the lines of communication between the two countries open and fostered commonality of interest. With the lessening of tensions after the end of the Cold War, opportunities for including Russia in international science projects and for joint U.S.-Russian space missions have increased. Supporting Russian efforts to maintain Russia's civilian space program could, however, help Russia become a stronger competitor to the United States.

Increased commercial ties between Russian and Western aerospace companies could also provide added incentive for Russia to abide by the MTCR. Russia has privatized several of its largest aerospace enterprises, which are now seeking customers for their products.

Fundamental Objective #3: The Use of Excess Ballistic Missiles

The third fundamental objective of the Clinton Administration's new space transportation policy addresses the use of long-range ballistic missiles that have been either superseded by more modern weapons or eliminated under the provisions of the Strategic Arms Reductions Talks (START) Treaty.

NSTP Fundamental Objective #3

Establishes policy on federal agencies' use of excess ballistic missile assets for space launch, to prevent adverse impacts on the U.S. commercial space launch industry. Under START, these assets may be used in certain circumstances for civilian space launch. A serious concern in developing the policy was the possible impact that widespread use of these assets could have on U.S. commercial launch companies. The policy obliges the government to fully consider commercial services as part of the decision making process and imposes specific criteria on the use of excess assets to avoid "flooding" the commercial market.

The policy language regarding excess missile assets reflects the consensus of an Interagency Working Group with representatives from DOD, NASA, DOT, DOC, the Department of State, the U.S. Arms Control and Disarmament Agency, and the USTR. The policy reflects the results of a subgroup jointly chaired by the National Security Council and OSTP.

The policy states that U.S. excess ballistic missiles shall either be retained for government use or be destroyed. DOD will consider whether to use excess ballistic missiles on a case-by-case basis,

[137] Since FY 1992, the Nunn-Lugar amendment to Public Law 102-228 and subsequent legislation have authorized the transfer of $1.6 billion of DOD funds to help destroy and secure weapons of mass destruction. Of that money, $25 million was to be the 1994 U.S. contribution to the International Science and Technology Center, which would provide research opportunities for FSU scientists in collaborative efforts with Western scientists. See U.S. Congress, Office of Technology Assessment, *Proliferation and the Former Soviet Union*, OTA-ISS-605 (Washington, DC: U.S. Government Printing Office, September 1994), pp. 23-28. Some U.S. private foundations have also made money available to Russian research institutions to try to curtail the proliferation of nuclear-weapons expertise.

[138] U.S. Congress, Office of Technology Assessment, op. cit., footnote 98.

and will require approval from the Secretary of Defense.[139]

The policy directs government departments and agencies requiring access to space to purchase commercially available U.S. space transportation products and services to the fullest extent feasible, and stipulates conditions on the use of excess ballistic missiles. Such use must support the sponsoring organization's mission; be consistent with international obligations, including the Missile Technology Control Regime (MTCR) guidelines and the START agreements; and result "in a cost savings to the U.S. Government relative to the use of available commercial launch services that would also meet mission requirements, including performance, schedule, and risk."[140]

In testimony before the Subcommittee on Space of the House Committee on Science, Space and Technology,[141] John Gibbons, Assistant to the President for Science and Technology, further clarified U.S. policy on use of missile assets, stating that "engineering tests and suborbital flight experiments are allowed, but orbital flights which may compete with private sector providers would have to satisfy some tough criteria."[142]

Gibbons argued that "these criteria are clear and reasonable and ... they provide sufficient flexibility to protect government interests while continuing to encourage private sector investment in new space transportation systems. If converting ballistic missiles to space launch vehicles can be done in a manner that saves money for the government, this policy will still allow us to take advantage of those savings."[143]

DOD does not currently plan to use intercontinental ballistic missile (ICBM) assets made ex-

U.S. DEPARTMENT OF DEFENSE

Static firing destroys a Pershing II engine under provisions of the Intermediate-Range Nuclear Forces treaty.

cess by START agreements for launching payloads into orbit, but its implementation plan notes that:

> . . .several contractors use ICBM heritage designs and tooling to produce new hardware similar or identical in design. New production of launch vehicle systems using ICBM technology and design will be allowed for either EELV competition or commercial application, but new productions must comply with existing policies of proliferation control of missile technology, as well as START provisions.[144]

Background

DOD currently has four U.S. strategic ballistic missiles in stock that are either retired or being retired. They are the Minuteman II and Titan II ICBMs, and the Trident I C4 and Poseidon C3 submarine launched ballistic missiles (SLBMs).

---

139 The White House, Office of Science and Technology Policy, op. cit., footnote 3, sec. VII(1).

140 Ibid., sec. VII(1)(c).

141 Now the House Science Committee's Subcommittee on Space and Aeronautics.

142 John H. Gibbons, Director, Office of Science and Technology Policy, "Statement on National Space Transportation Policy," testimony at hearings before the Subcommittee on Space, Committee on Science, Space and Technology, U.S. House of Representatives, Washington, DC, Sept. 20, 1994.

143 Ibid.

144 U.S. Department of Defense, op. cit., footnote 125.

BOX 9: Treaty-Driven Force Reductions

The first Strategic Arms Reduction Treaty (START I), signed by President Bush and Soviet President Gorbachev on July 31, 1991, called for limiting each nation to 6,000 accountable warheads on 1,600 strategic offensive delivery vehicles.

As a result of START I, the United States has removed warheads from all 450 single-warhead Minuteman II ICBMs. In addition, the United States has removed the missiles from all its Poseidon ballistic missile submarines. Nearly half of the missiles from 31 Poseidon submarines that were in the U.S. fleet have been eliminated.

Protocols to the START I treaty now in force call for the elimination of most surplus ballistic missiles. The Nineteenth Agreed Statement annexed to the START I agreement requires that a party to the treaty wishing to use surplus missiles to develop space launch boosters must get the consent of the Joint Compliance and Inspection Commission. Converted space transportation systems would be allowed provided the resulting boosters differ verifiably from ICBMs and SLBMs, and provided the number of boosters produced and stored do not exceed space launch requirements. START I might result in a surplus of as many as 450 ICBMs and 192 SLBMs.

START II was signed on January 3, 1992, but has not been approved by either the United States or Russia. This treaty would further reduce warheads on strategic offensive delivery vehicles to 3,500 or fewer and would ban multiple warhead ICBMs. Russian ratification appears uncertain.[1] One U.S. deployment option under START II could add 50 Peacekeeper ICBMs to the surplus created by START I. START II also calls for reducing the number of warheads on SLBMs from eight to four, but without reducing the number of missiles themselves.

[1] "Russian Parliament Approval of START II 'Uncertain at Best': CIA Aide," *Aerospace Daily* 173(40):311, 1995.

SOURCE: Office of Technology Assessment, 1995.

Some of the retirements are a result of various arms control agreements (see box 9).

Minuteman

All of the 450 single-warhead Minuteman IIs were taken off alert in September 1991 and will be retired under START.[145] By the end of 1994, the Air Force had removed 384 Minuteman II ICBMs from their silos and had transported them to storage or to processing facilities. The United States plans to deactivate the remaining 66 Minuteman IIs by the end of 1995. Martin Marietta has advocated refurbishing the Minuteman II for use as a small launcher capable of supporting suborbital experiments and carrying 1,200-lb satellites to LEO. Minuteman IIs have been used for several Strategic Defense Initiative tests and were once considered for use in evaluating National Aerospace Plane (NASP) technologies.[146] Furthermore, a Minuteman II was an early booster choice of the Universities Space Research Association (USRA) to launch scientific payloads in its Student Explorer Demonstration Initiative.

[145] The Arms Control Association, "U.S. and Commonwealth MIRVed Strategic Ballistic Missiles: Fact Sheet," Washington, DC, Jan. 10, 1991.

[146] Leonard David, "NASP Backers Seek Scramjet Tests on Surplus Missiles," *Space News*, Mar. 22, 1993.

Titan II

The Air Force has retired some 53 Titan II boosters from silos in Nebraska, Kansas, and Wyoming.[147] The Air Force is currently evaluating how many of these missiles to retain for other uses and how many to destroy. Engines from the Titans are likely to be returned to Martin Marietta Denver for retrofitting into current Titan IIG SLVs. At least two firms are believed to have contacted the Air Force stating their intentions to bid on the Med-Lite competition using Titan IIs, but the bids never materialized.

Martin Marietta has an Air Force Space Command contract to refurbish retired Titan IIs for use as SLVs. The contract calls for the conversion of 14 Titan II missiles into Titan IIG SLVs, to be completed by September 1995. As a result of that contract, the Ballistic Missile Defense Organization was able to boost its Clementine lunar spacecraft into orbit on a refurbished Titan IIG in January 1994.

Trident/Poseidon

The United States plans to retire 92 Trident I C4 missiles when their submarine carriers are retired. The current schedule calls for the destruction of these SLBMs after Lockheed Missiles and Space Co. evaluates whether they might have commercial use. Lockheed has expressed an interest in using the C4s for sounding rockets or small satellite boosters. Navy officials, however, claim that the C4s will ultimately be destroyed. The Navy also plans to destroy the 160 Poseidon C3 missiles that were retired as of the end of 1991.

Issues for Congress

Arms control agreements and military downsizing have left both the United States and Russia with a significant surplus of long-range ballistic missiles. Both countries have the option to convert them to space launch vehicles, but, to date, the United States has generally decided to forgo this option. The Russians, on the other hand, are pursuing several conversion projects. Both of these choices raise issues for Congress.

U.S. DEPARTMENT OF DEFENSE

A Titan II missile silo in Arizona is destroyed under terms of the Strategic Arms Limitation Treaty II.

ISSUE 3a: *Unfair competition or market creation?*

Before the Clinton Administration's policy on use of missile assets, former Vice President Quayle, while head of the National Space Council, sponsored a set of studies on the future of America's space capabilities. The Vice President's Space Policy Advisory Board issued a report on U.S. space launch capability that called attention to the promise of using excess ballistic missiles.[148] At the time, the Bush Administration had not taken a formal position on the use of excess ballistic missiles for commercial space launch, but in the interim had denied the use of

147 David Mosher, Congressional Budget Office, private communications, Mar. 22, 1995.

148 Vice President's Space Policy Advisory Board, "The Future of the U.S. Space Launch Capability: A Task Group Report," November 1992, pp. 7-8, 26, 35-36.

TABLE 14: Use of Excess Ballistic Missiles

Benefits	Costs
Lowers cost of access to space. Increases space-related R&D in both the commercial and academic sectors. Creates business opportunities to convert surplus missiles and to provide associated launch services. Tests the market for SLVs, which will allow entrepreneurial firms to raise investment capital for the creation of new, more competitive launch vehicles and services.	Undermines the commercial production of space launch vehicles by reducing their volume, thereby raising their cost. Stalls the development of new, more efficient SLVs.

SOURCE: Vice President's Space Policy Advisory Board, "The Future of the U.S. Space Launch Capability: A Task Group Report," November 1992.

these assets pending completion of a review.[149] The report's review of potential benefits and costs of converting Titan II, Poseidon, and Minuteman assets is summarized in table 14.

The task group suggested that a balance between the two points of view must be found, and recommended the establishment of a government-supported, small payload launch program. This program would use low-cost launch vehicles to promote and encourage space research and experimentation and permit the use of excess ballistic missiles as space launch vehicles for government-sponsored research or commercial applications under specifically controlled conditions.

Most of the points raised by former Vice President Quayle's task force continue to be valid. At a time when the United States is trying to reduces its budget, however, the requirement for less-expensive launch vehicles is paramount, not only for military and scientific purposes, but for commercial needs as well.

Recycling missile assets has been viewed by some as a means of reestablishing U.S. leadership in commercial space transportation. Surplus missiles could be retrofitted to accommodate the growing requirement for low-cost launch vehicles. In the process, the government would be saved the expense of both scrapping the missiles and buying new SLVs. Over 600 ballistic missiles are slated for retirement by the end of the decade in the United States alone.[150]

A lucrative market is now evolving to provide a new generation of global satellite telecommunications.[151] Private firms are developing constellations of small satellites that are scheduled to be placed in LEO in the near future. The number of spacecraft required for each constellation, and the need to replenish individual satellites as they deteriorate, is adding to an appetite for low-cost launch vehicles. Moreover, the compact nature of these spacecraft permits the use of smaller launch vehicles.

Still, many U.S. commercial launch providers view the release of retired missile assets as a threat to their industry. They believe these assets will not be priced according to their true costs. Moreover, they point out that the United States is already the dominant provider of commercial SLVs. Converting excess ballistic missiles might lower government costs and expand the U.S. share of the world market in the near term, but could inhibit long-term investment in SLV development.

[149] Ibid., p. 26.

[150] David Mosher, Congressional Budget Office, private communications, Mar. 22, 1995.

[151] See the section covering fundamental objective #4.

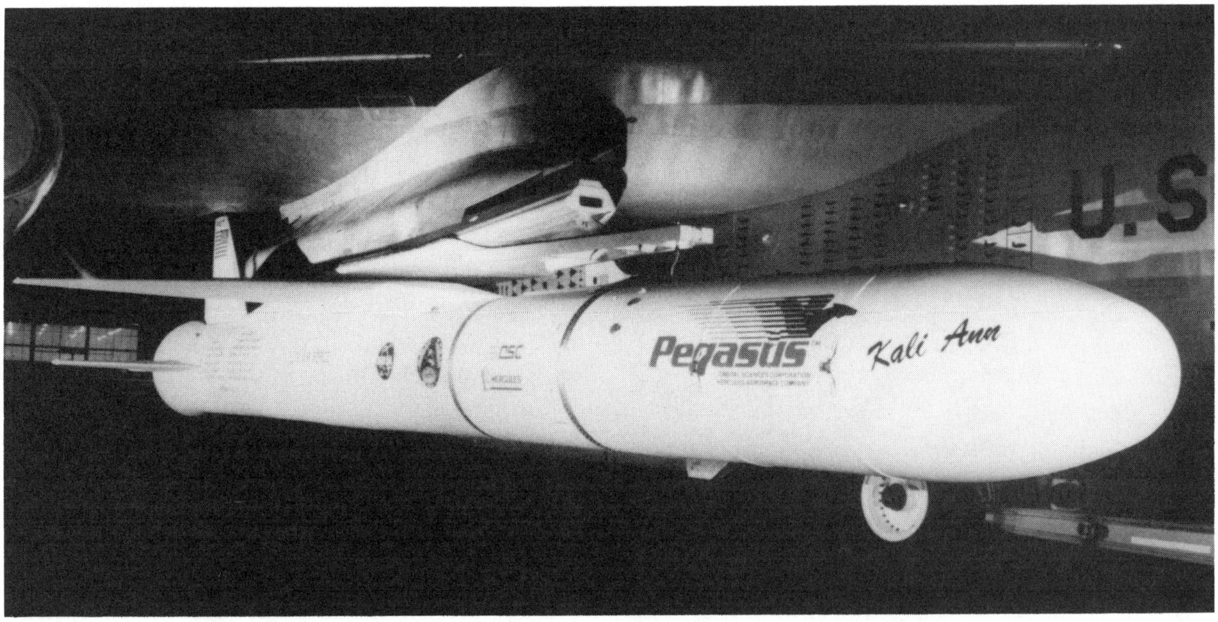

ORBITAL SCIENCES

Converted excess ballistic missile assets would compete with existing U.S. SLVs, such as this Pegasus.

The availability and possible use of retired missile assets in the United States has been hotly debated, most ardently between the Orbital Sciences Corp. (OSC) of Dulles, Virginia, and officials at USRA. OSC opposes use of missile assets for orbital insertion of payloads while USRA favors using the less-expensive assets, in particular the Minuteman II, to launch small scientific payloads.[152]

As a practical matter, DOD and NASA have generally discouraged the conversion of ballistic missiles. DOD did offer NASA use of its Titan II for NASA's Med-Lite missions, but NASA decided instead to purchase the services commercially.[153]

Some companies have opted to avoid the issue of excess ballistic missiles, focusing instead on adapting ballistic missile production technology and designs to new commercial vehicles. For example, on January 23, 1995, E'Prime Aerospace Corp. of Titusville, Florida announced that it had received DOD approval to market a space launch vehicle based on Peacekeeper ICBM technology. E'Prime claims that, under an amendment to its Commercialization Agreement with the U.S. Air Force, it has the rights to use Peacekeeper technology to develop a commercial launch vehicle—the Eagle S Series. No existing Peacekeepers themselves will be used, but Peacekeeper tooling and ground support equipment has been purchased from the government.[154]

Lockheed Missiles and Space Co. offered to use a modified Poseidon C3 to launch small satellites for the military several years ago, but "vague and uncertain DOD policies" did not warrant de-

---

[152] OSC supports the conversion of ballistic missiles for suborbital launches. Such a policy poses no threat to OSC's standing in the SLV market.

[153] For a detailed discussion of this decision, see box 6.

[154] Bob Davies, President, E'Prime Aerospace Corp., private communications, April 1995.

LOCKHEED MARTIN

Lockheed privately financed development of the LLV.

velopment of the concept, according to Lockheed officials.[155] However, the effort led to the company's development of a family of Lockheed Launch Vehicles (LLV). LLV development has drawn on tested and tried motors, as well as other off-the-shelf components. The maiden flight of an LLV-1 is slated for flight from the Vandenberg Air Force Base in California sometime in 1995.

Given the restricted budgets available within the government to pursue new types of launch vehicles, the prospect of falling back on surplus missiles for certain smaller classes of payloads might be attractive, particularly if these missiles could be used to create a market for future SLVs. The cost of using these missiles, however, might be too high if it undermines the ability of commercial SLV providers to develop new launch vehicles and, perhaps, even drives them out of the space transportation business.

ISSUE 3b: *Russian excess ballistic missiles*

Although in the United States the use of excess ballistic missiles is being tightly controlled, some Russian enterprises are promoting a number of converted ICBMs and SLBMs for an assortment of commercial uses.[156] The Scientific and Technological Center (STC) of Moscow has begun to market surplus SS-25 ICBMs, rebuilt to launch satellites. Called the Start-1, the vehicle made a demonstration flight from Russia's Plesetsk Cosmodrome on March 25, 1993.

The Start-1 is a transportable, four-stage booster derived from a road-mobile, solid-propellant ICBM, capable of launching small microgravity payloads or small LEO satellites. Last year, CTA, Inc. of Rockville, Maryland announced that the Start-1 was the firm's vehicle of choice to place into orbit a commercial remote sensing satellite built by CTA for EarthWatch, Inc.[157] According to CTA, the Russians offer cheaper SLVs with preparation times faster than anything currently available in the United States. CTA's decision to use the Start-1 could change if the Russians do not

---

[155] Howard Trudeau, Vice President, Engineering, Missile Systems Division, Lockheed Missiles and Space Co., "The Lockheed Launch Vehicle Family," *Sixteenth Annual Lockheed Technology Symposium*, Washington, DC, Nov. 16, 1994, p. 2.

[156] U.S. Department of Transportation, Office of Commercial Space Transportation, Commercial Space Transportation Advisory Committee (COMSTAC), "Report of the COMSTAC Task Group on Soviet Entry Into the World Space Market," Washington, DC, August 1992.

[157] Boris Feldblyum, "Peace Dividends—Refurbished Russian ICBM To Carry U.S. Imaging Payload," *The International Small Satellite Organization Newsletter* 8(2):5-6, April/May 1994.

adequately analyze and explain recent failure of a Start-1 fifth stage.[158]

The Russians have converted another ballistic missile, the SS-19, into a space launch vehicle called "Rokot." This vehicle boosted an amateur radio satellite into orbit on December 26, 1994. The Rokot is a liquid-fueled ICBM developed by the Salyut design bureau in Moscow, now under the auspices of the Khrunichev Space Research and Producing Association. An upper stage of this converted missile, however, blew up after placing its payload into orbit.[159]

In addition, the Russians are promoting conversion of the SS-18, the most powerful ICBM (largest throw-weight) developed in either the FSU or the United States. This two-stage, liquid-fueled booster could easily provide LEO launch services. Australian space officials and Russian rocket producers from STC Complex have discussed converting SS-20 and SS-25 missile stages, modified with a more accurate guidance system, into SLVs.[160] The resulting vehicle would launch satellites from Australian launch sites near the equator, which would allow the SLVs to reach orbit using minimum fuel.

Russian companies have also unveiled plans to develop launch vehicles derived from a number of Russian SLBMs. The SS-N-8, -18, and -23 liquid-fueled rockets have been touted as capable of lofting microgravity capsules based upon warhead reentry vehicle designs.

Finally, a Ukrainian venture begun in 1990 and led by Scientific Production Organization (NPO) Yuzhnoye, in conjunction with NPOs Soyuz and Iskra, converted the Ukrainian SS-24 into an air-launched vehicle called "Space Clipper."[161] Converting this vehicle means retaining its lower three solid-fueled stages, and developing a new fourth stage and control system.

The availability of this wide assortment of commercial launch assets, all based on converted Cold War missiles, requires development and production money, as well as commercial customers. Whether or not these military missiles of the FSU will evolve to true commercial status remains to be seen. Nonetheless, if the U.S. government believes that surplus U.S. ballistic missiles undermine our domestic SLV launch providers, then Russian ballistic missiles may be seen as an equal, or greater, threat.

Fundamental Objective #4: The Private Sector Role in Space Transportation Decisionmaking

The NSTP's fourth fundamental objective stresses expansion of the role of the private sector in space transportation R&D in order to meet the government's need for assured access to space at an affordable price and improve the international competitiveness of the U.S. private-sector space transportation industry.

Fundamental Objective #4

Provides for an expanded private sector role in the federal space transportation R&D decision making process. In contrast with previous national policy on space transportation, this policy specifically directs the Departments of Transportation and Commerce to identify opportunities for government-industry cooperation and to factor these into NASA's and DOD's implementation plans.

The private sector has been included as a critical element of U.S. space transportation policy for

[158] Sharone Parnes, "Israelis Regroup After Loss of Satellite on Russian Launcher," *Space News*, Apr. 3, 1995, p. 38; and private communications with a representative of CTA, Inc.

[159] "Converted Missile Explodes After Launch," *Space News*, Feb. 6, 1995, p. 2.

[160] Reuters Ltd. Wire Service, "Australia Eyes Russian Missiles for Space Industry," from Canberra, Australia, Feb. 8, 1995.

[161] Chris Bulloch, "Destroy Them or Launch Them?" *Interavia*, January 1995, pp. 44-47. The Space Clipper is not carried outside the aircraft as is the OSC's Pegasus, but is extracted from the rear cargo doors of the Russian AN-124 heavy transport plane.

ROCKWELL INTERNATIONAL

NASA hopes to replace the Space Shuttle with a commercially owned and operated RLV, or perhaps privatize the Shuttle.

several years. One reason often cited for the commercial emphasis is the desire to preserve an important high-technology, commercial industrial sector in the United States. But a more compelling reason for the current emphasis on the private sector appears to be the judgment that, for the foreseeable future, budget constraints make it unlikely that the government will pay the entire cost of developing and maintaining significantly new national space transportation capabilities. In this fiscally constrained environment, private sector financing is viewed by many observers as essential for NASA's development of a follow-on to the Space Shuttle.[162]

Further, there is a perception that the government should have less responsibility in areas in which the private sector might reasonably be expected to provide the desired goods and services. The privatization of most launch facilities and many of the actual launch activities, for example, is often considered by many to be necessary if the United States is to have a viable space transportation industry in the future. Greater use of the private sector conforms with current DOD and National Performance Review acquisition policy initiatives that stress increased use of commercial products and processes. Under these circumstances, increased private sector involvement in the R&D decisionmaking process appears not only prudent, but absolutely essential.

Nevertheless, the past experience of government research programs has caused some observers to question the potential for the private sector to have meaningful input into the space transportation development process. Discussions with industry raise significant questions about the conditions under which the private sector (which must remain profitable to survive) would be willing to finance the development and building of a new space launch vehicle.

This section outlines government policy and implementation plans related to an expanded role for the private sector, discusses the potential of the implementation plans to meet stated government goals for the commercial sector, and identifies issues of potential concern relating to the current policy and implementation plans.

Government-Industry Goals and Policy

The U.S. government space transportation goal of assuring reliable and affordable access to space using domestic capabilities subsumes a commitment to a viable and internationally competitive U.S. commercial space transportation industry. Indeed, the policy directs government entities to "encourage the cost-effective use of commercially provided U.S. products and services, to the fullest extent feasible, that meet mission requirements; . . . [and] . . . foster the international competitive-

162 Ivan Bekey, NASA, "A Win-Win Concept for Commercial Development and Operation of a New, Large Reusable Space Launch Vehicle: An "Existence Proof" White Paper," Dec. 21, 1994, pp. 2-3. Bekey reports that NASA concluded that the government was very unlikely to proceed into procurement of a new launch system in the future, and that both NASA and the DOD independently concluded that if the next generation reusable launch vehicle were to be developed, it would have to be done by the private sector using largely private sector funding.

ness of the U.S. commercial space transportation industry, actively considering commercial needs and factoring them into decisions on improvements in launch facilities and launch vehicles."[163]

Recognizing the need to design for dual or multiple use to enhance U.S. commercial competitiveness, the policy directs government planners to involve the private sector in the design and development of space transportation capabilities; transfer unclassified, government-developed, space transportation information to industry in a timely and commercially relevant manner; and promote common technical standards for space products and services.

The policy further directs government planners to help identify and promote innovative ways for the private sector and federal, state, and local governments to work together to implement space transportation policy; to avoid engaging in activities that have commercial applications and that might deter commercial space activities; and to provide industry stable and predictable access to appropriate space transportation hardware, facilities, and services.

The Implementation Plans

The implementation plans all specifically address the role of the private sector in meeting their assigned goals, but organizational views on this issue—as expressed in the policy, plans, and funding levels—vary considerably.

Departments of Transportation and Commerce Implementation Draft Plan

The DOT/DOC implementation plan was still under revision during the writing of this report. However, the latest available draft advances an objective for the U.S. space transportation industry to "capture a dominant portion" of the global market for launch services by encouraging the development of a more internationally competitive launch vehicle fleet and supporting infrastructure.[164]

The importance and difficulty of obtaining financing (private or public) in the space launch industry has been a matter of concern for several years. The NSTP gives DOT and DOC specific responsibility for identifying opportunities for government-industry cooperation and for promoting innovative types of arrangements to implement the policy. The DOT/DOC implementation plan lists, and briefly discusses, several options for stimulating private investment in space transportation vehicle development and infrastructure improvements.[165] Some of these options are included in table 15. The potential benefits and drawbacks of the most important of these options are discussed later in the issues section.

Industry has expressed particular interest in **anchor tenancy** and termination liability. The private sector and the government frequently use anchor tenancy agreements to support financing of new building construction. Such agreements allow the building developer to raise funds to construct the building. Early leases can result in lower rates or other benefits for the anchor tenant.

Similarly, the government might act as an anchor tenant for a space transportation system by providing a guaranteed launch market for a specific period of time to the space transportation provider. As an anchor tenant, the government would provide an income stream and reduce the investment risk for the private sector during a period in which commercial markets are being established. For example, the government might negotiate to purchase a designated amount of lift capacity (with some specified performance and schedule minimums) per year to LEO, to make a 10-year

---

[163] The White House, Office of Science and Technology Policy, op. cit., footnote 3, secs. I(5) and I(6).

[164] U.S. Department of Transportation/U.S. Department of Commerce, "Department of Transportation/Department of Commerce National Space Transportation Policy Implementation Plan: Executive Summary," Washington, DC, Apr. 19, 1995 (draft).

[165] Ibid., pp. 8-11.

TABLE 15: Options for Financing Launch Vehicle Development

General option	Examples
Anchor tenancy	Government commitment to procure a specified number or percentage of goods and/or services over a number of years.
Termination liability	Government commitment to reimburse a contractor for all or part of its investment if the government terminates a contract for its convenience.
Public/private partnerships	R&D limited partnerships, in which investors could form a limited partnership to conduct cooperative R&D; consortia; space transportation corporations; space service brokerages; contracted partnerships; and government prizes.
Other financial options	Loan guarantees, tax deferments, exemptions, and credits.

SOURCES: DOT/DOC, "Department of Transportation/Department of Commerce National Space Transportation Policy Implementation Plan" (Draft), Nov. 7, 1994, pp. 9-12. H.R. 6135, *NASA FY 93 Authorization* (Public Law 102-588). H.R. 258, *Launch Services Corporation Act of 1995*.

purchase commitment, and to have some potential for increasing the commitment.[166]

Termination liability commits the government to provide payment to a contractor in the event the government terminates a contract for its convenience. Advocates argue that, when applied to the space transportation industry, this arrangement would protect private sector firms involved in endeavors such as multiyear anchor tenancy against the government's terminating a contract after a firm had made large investments.

A **space transportation corporation** might be established to provide economies of scale to the U.S. space transportation industry by pooling all government launches as a central procurement agent. It might be quasi-public, chartered by Congress, and modeled on COMSAT.[167] The corporation would work to increase the economic exploitation of space and enhance the economic competitiveness of the U.S. space transportation industry. The corporation might initially have some government capital and operate with an anchor tenancy arrangement. Whatever the structure, proponents argue that a corporation would operate on business principles and use business practices.

Activities such as **R&D limited partnerships** and **consortia** would be aimed at leveraging investments by individual firms as well as the government. Such activities are currently sanctioned by the National Cooperative Research Act of 1984.[168] Joint R&D has been used extensively in recent years in many fields.

The DOT/DOC plan also includes the idea of offering a **government prize** for the development of a good or service the government desires. This concept has been advocated by some in industry as an efficient way to entice private sector investment in space launch. Under this concept, the government might simply award a prize to the first competitor meeting the government's perfor-

[166] See T.F. Rogers, "Toward a New Public-Private Space Transportation Strategy," *The Journal of Practical Applications in Space* 5(1):17, 1993. Ivan Bekey, op. cit., footnote 162, suggests a five-year, above-market price guarantee to help industry secure initial financing. Industry representatives provided OTA with a number of alternative anchor tenancy scenarios.

[167] For one possible structure, see U.S. Congress, House of Representatives, H.R. 258, "Launch Services Corporation Act of 1995," Washington, DC, 1995 (introduced on Jan. 4).

[168] U.S. Congress, Office of Technology Assessment, *Making Things Better: Competing in Manufacturing*, OTA-ITE-443 (Washington, DC: U.S. Government Printing Office, February 1990), pp. 225-226. The National Cooperative Production Amendments of 1993 amended the National Cooperative Research Act of 1984 to include joint production activities.

mance needs (e.g., a vehicle that could deliver a 25,000 lbs of payload to the International Space Station, have a total launch operations crew of fewer than 50 people, and take less than four days after return to be ready to launch on a new mission). A sufficiently large prize, some advocates argue, would be a far greater stimulus of innovative developments and new entries into the launch business than anchor tenancy or termination liability.

Loan guarantees, tax deferments, exemptions, and credits are among the other financial options listed in the DOT/DOC implementation plan. The Japanese government, for example, has used success-dependent loans (*hojokin*) to assist risky enterprises in selected industries. Repayment for these loans can wait for a positive cash flow.[169] Tax credits and exemptions are more traditional U.S. policies. DOD has employed Title III of the Defense Production Act to develop production capabilities in areas deemed essential to national security. DPA money, for example, is currently being used to fund part of the government's Flat Panel Display Initiative.[170]

The DOT/DOC implementation plan also suggests that significant economic savings might be realized through the development of joint-use facilities. Many observers believe that important savings are to be gained from changes in launch operations and facilities. The DOD-administered, dual-use space launch facility program has funded studies of commercial spaceports, initially using surplus government facilities. This is discussed in more detail later.

The DOT/DOC implementation plan notes that the principal role for federal funds may be as "the provider of matching grant funds, venture capital, credit capacity, and a 'bankable' stream of revenue to support commercial launches." This view of the government as the ultimate source of funding is a more traditional view of government's relationship to the launch industry. It is consistent with the findings of the 1994 "Commercial Space Transportation Study" (CSTS), whose participants reported they were unable to prove that the foreseeable commercial space market was sufficiently robust to support a completely commercially developed system. The study therefore concluded that some level of government financial participation was essential to attract commercial investment in U.S. space transportation.[171]

Department of Defense

The DOD implementation plan includes the private sector less as a partner in the development of new, commercially applicable launch capabilities, than as a provider of less costly launch vehicles to meet DOD needs. The Department supports U.S. commercial programs indirectly through its EELV development and procurements. U.S. firms are expected to derive a spinoff benefit from the development of a lower cost EELV that might then be offered to the commercial market.[172] The Department's operation and maintenance of much of

[169] U.S. Congress, Office of Technology Assessment, *Competing Economies: American, Europe and the Pacific Rim*, OTA-ITE-498(Washington, DC: U.S. Government Printing Office, October 1991), p. 68.

[170] Title III of the Defense Production Act of 1950 (50 U.S.C. App. 2061) addresses the expansion of productive capacity and supply. Under this authority, the President may make purchases, guarantee purchases or guarantee loans to help develop a capability that is viewed as essential for national security.

[171] The Commercial Space Transportation Study Alliance, op. cit., footnote 16, p. 13. The CSTS is an examination of potential markets for space transportation undertaken by the CSTS Alliance. The Alliance consisted of six firms: Boeing, General Dynamics, Lockheed, Martin Marietta, McDonnell Douglas, and Rockwell. The study began in March 1993 and reported findings in April 1994.

[172] U.S. Department of Defense, op. cit., footnote 125, p. 8.

the national space launch infrastructure is a benefit to the private sector.

Moreover, DOD reports that it is making efforts to design, or redesign its payloads to have more "commercial" characteristics.[173] One of the principal problems, however, is that the national security community retains a requirement for a heavy-lift capability that is very costly. That capability is not currently of great use in the commercial market.

The DOD implementation plan incorporates many of the recent DOD acquisition reforms aimed at opening the Department to commercial developments and commercial business practices. Performance specifications and commercial specifications and standards will be used in lieu of military specifications and standards, for example, unless there are no practical alternatives. Furthermore, the plan encourages industry to offer alternative solutions to the total government mission model requirements, rather than to meet detailed government specifications.[174] As a part of these changes, foreign components may also be used, "but sole dependence on foreign sources of supply will not be permitted."[175]

The DOD plan encourages commercial investment, but with some caution. The plan directs that if investment is included, bidders "must identify how they intend to recover their investment in the EELV recurring cost."[176]

NASA

The NSTP directs NASA to actively involve the private sector in planning and evaluating its launch technology activities. NASA's implementation plan envisions a "government-industry partnership" aimed at developing and demonstrating new reusable space transportation technologies. The Agency believes that these technologies have the potential to reduce the cost of access to space radically and act as a catalyst for the increased use of space.

NASA reports that a government-industry partnership is essential, because:

> . . . the private sector could have a significant role in managing the development and operation of a new reusable space transportation system. In anticipation of this role, NASA shall actively involve the private sector in planning and evaluating its launch technology activities.[177]

NASA's program objectives for the RLV development program include the reduction of technical risks associated with building and operating a reusable system in order to encourage private investment in the commercial development and operation of the next-generation system.[178] The Agency's plan describes a concept in which industry might make significant decisions, and "involvement by the government, at industry's request, will be in areas where the government's technical expertise and assets can be used to their fullest advantage."[179]

NASA's implementation plan describes joint "government-industry synergy teams" that select key technologies, design test programs, and develop evaluation criteria for validating the technology in an integrated system with realistic operations, maintenance, and flight environ-

[173] U.S. Department of Defense, Office of the Under Secretary of Defense, Acquisition and Technology, "MED-LITE: A DOD White Paper," Washington, DC, December 1994. This was discussed in detail earlier in the section on fundamental objective #1.

[174] U.S. Department of Defense, op. cit., footnote 125, p. 7.

[175] Ibid., p. 6. This is discussed earlier in the section covering fundamental objective #2.

[176] Ibid. Foreign sourcing of components will be managed so that foreign suppliers cannot deny items to the United States, using stockpiling and assuring that alternative sources of supply can be developed.

[177] NASA, op. cit., footnote 29, p. 3. Some skeptics note that the implementation of any partnership may well be challenged by the fact that NASA has man-rated space launch requirements, while the commercial industry does not generally need man-rated vehicles.

[178] Ibid., p. 5.

[179] Ibid., p. 4.

ments. The Agency believes such a program will provide practical experience in routine operations of reusable systems.[180]

Issues for Congress

The private sector policy goals and the measures outlined to achieve them in the implementation plans raise a number of critical issues involving the size and nature of the future space transportation market; the nature of the government-industry relationship; the proper balance between risk and incentives; launch operations and infrastructure; and the willingness of the government to accommodate commercial needs in seeking to achieve its own goals. These issues are discussed below.

ISSUE 4a: *Will the estimated market support policy goals?*

The assessment of the size, character, availability, and relationship (e.g., potential overlap) of future space transportation markets is critical to industry's attitude toward new space transportation development programs and government cost-sharing schemes.

Industry sees several different market segments: U.S. government (defense, intelligence, and civil), U.S. commercial, foreign government, and foreign commercial. Assuming no significant near-term change in the United States' national security situation, a core market, composed of announced government and commercial programs, is relatively well defined. There is, however, considerable uncertainty about the size and availability of the overall potential space transportation market. The CSTS, for example, estimated a significant increase in the use of space, and therefore in commercial launchings—if the cost of putting payloads into orbit were greatly reduced. But many of the increased uses of space envisioned in the CSTS were expected to come in nontraditional

HUGHES AIRCRAFT

Satellites, such as this HS 601, provide capabilities for air traffic control, mobil phone service, and television, voice, facsimile, and data transmission.

areas (e.g., tourism and waste disposal). Indeed, the study acknowledged that currently projected markets are insufficient to attract major investments, and argued that "to become economically viable, a new launch system must generate new commercial markets."[181]

The current industry assessment appears to be that:

- The potential commercial market for MLVs is *insufficient by itself* to entice enough private investment to build a future RLV capable of meetings NASA's needs.
- The potential commercial market for small payloads *may be sufficient* to attract enough private investment to develop vehicles to meet both commercial and government needs for small payload delivery.

---

[180] Ibid., p. 6.

[181] The Commercial Space Transportation Study Alliance, op. cit., footnote 16, p. 1.

U.S. DEPARTMENT OF DEFENSE

A Navstar Global Positioning Satellite is prepared for launch aboard a Delta II.

- There is little estimated commercial heavy-lift market and the private sector is unlikely to put much of its own funds in that area without substantial government support.

Such assessments imply that if the government wishes to encourage the development of a commercial RLV, it will have to provide a significant amount of funding—either through a direct development and procurement process or through some form of guaranteed business.

Estimating the Market

Government and industry planners use "mission models" that contain estimates of future space transportation needs to help them make investment decisions. These models may vary depending on the developer's judgment about the likelihood of a particular action occurring (e.g., the deployment of one of a number of proposed global satellite communications systems).

Because of the time involved in developing new space transportation systems and the large investments involved, the investment time periods of interest are necessarily long—usually 15 to 20 years. Market projections become progressively ill-defined further in the future. The uncertainties of these markets raise the financial risks for the private sector. These higher perceived risks, in turn, cause firms to seek risk mitigation alternatives such as anchor tenancy, termination liability, and other arrangements designed to hedge against inadequate financial returns.

The U.S. Defense and Intelligence Market

DOD payloads have been a major component of the global space launch market for U.S. space transportation firms. Under any scenario, they will continue to be important in the foreseeable future. In the absence of a new, major, military threat, or a change in defense planning, estimates of the U.S. national security market for the next decade are fairly well defined.[182] The DOD recently estimated a demand for an average of about two small, eight medium, and three heavy launches per year between 1995 and 2010.[183] As noted earlier, the heavy launches consume roughly 80 percent of the Department's space transportation budget and are its principal target for EELV savings. (See section on fundamental objective #1.)

[182] Major new deployments such as a ballistic missile defense could have a significant effect on these estimates.

[183] U.S. Department of Defense, op. cit., footnote 15, p. II-2. Estimates for commercial medium-class launches have recently increased slightly, reducing the government's percentage of that important market segment. In 1994, DOD launched 5 small payloads, 4 medium payloads, and 4 heavy payloads. These 13 launches constituted 48 percent of the 27 total U.S. launches. There was a worldwide a total of 94 space launches in 1994. In addition to the United States; Russia launched 49; France launched 8; China, 5; India, 2; and Japan, 2.

DOD estimated that government launches will likely comprise almost all the of the U.S. HLV vehicle market, from 30 to 40 percent of the U.S. market for MLVs (depending on commercial communications developments), and from 15 to 30 percent of the U.S. SLV.[184] Most industry estimates agree on the importance of the DOD market to any U.S. space transportation business strategy. DOD currently provides the principal U.S. ELV market and is potentially an important component of the future RLV market.

The Government's Nondefense Market

The U.S. government's civil space transportation market is composed of weather and communications satellites; deployment and support of the International Space Station; and other planned scientific missions. NASA relies primarily on the Space Shuttle for its manned and scientific missions, but it also uses ELVs for some additional scientific, communications, and Earth observation missions. The National Oceanic and Atmospheric Administration relies on ELVs. DOD's industrial base study estimated the government's nondefense launch market at 18 missions per year (three small, seven medium, and eight Shuttle). NASA has estimated a need for about 15 missions per year, principally MLVs and the Shuttle.[185]

Industry planning estimates for the civil market are generally conservative and are close to these government figures, although they agree with the findings of the CSTS that significantly lower launch costs would probably expand current market segments as well as establish new ones. An important issue, however, is how much and at what rate expansion might occur. In a fiscally constrained environment, the government's nondefense market for space transportation may not be very responsive (at least in terms of dollars spent) to reduced launch costs.

The Commercial Market

The size and character of the future commercial market are critical to private business as well as to government space transportation decisions. How large this market segment might grow, and how fast that growth might occur, will decide the success or failure of many of the government's policies involving industry participation. This market segment is, however, the most difficult to predict. Private sector interest in government guarantees, such as anchor tenancy and termination liability, indicates a broad skepticism about the size of near-term commercial markets. An exception already noted may be the small payload (under 2,000 lbs to LEO) sector, which is widely viewed as having considerable growth potential.

Since 1990, U.S. launch providers and Arianespace have launched an average of 23 commercial satellites each year (see table 16). The draft DOT/DOC implementation plan estimates a future annual launch rate for the international commercial market of medium-to-large communications satellites in the range of 12 to 15. NASA's "Access to Space Study" estimated a range of 5 to 14 commercial launches per year through the year 2030. DOD estimated an average of 15 per year through 2010, excluding recently proposed LEO satellites. The LEO constellations might be expected to add 4 to 10 medium-to-large launches per year during the currently estimated deployment phases (1996-98 and 2002-2005), depending on what systems are deployed, and some 8 to 12 small vehicle launches per year.[186]

---

[184] U.S. Department of Defense, op. cit., footnote 15, p. II-3.

[185] U.S. Congress, Office of Technology Assessment, op. cit., footnote 87. These include 2 Pegasus/Taurus Class, 3 Delta Class. 2 Atlas Class, 0.3 Titan IV Class, and 8 Shuttle, p. 5. There were seven Shuttle and three ELV NASA missions in 1994.

[186] Frank C. Weaver, Director, Office of Commercial Space Transportation, U.S. Department of Transportation, "Dear Colleague," letter to industry addressing estimated LEO launch services, Mar. 13, 1994.

ARIANESPACE

Arianespace developed the heavy-lift Ariane 5 to launch more than one small-or medium-sized payload at a time.

Industry estimates are generally of the same magnitude, although industry points out the negative effects that the variability of the market can have on business. Several communications satellites sent aloft about the same time as a part of a communications network, for example, will need replacement at about the same time, causing anticipated peaks and valleys in the launch business.

Industry is starting to include in its commercial market forecasts some proposed LEO communication satellite constellations. The proposals include constellations that range from 12 to 840 or more satellites (see table 17). These constellations will provide cellular communications from space.[187] The more mature LEO communications concepts (e.g., Iridium) are going forward in the absence of new and cheaper launch systems.

Commercial launch rate estimates are probably fairly accurate out to about five years since they are principally based on payloads already in production or scheduled for production. Further into the future, however, commercial launch needs become much more uncertain and affected by new technological developments and new areas for market growth. From the U.S. industry viewpoint, the market will also be affected by the number of launch system providers—the United States, Europe, Russia, Japan, China, and others.

The CSTS estimated that the number of launches, and the total mass in orbit, would increase sharply if launch costs (and launch prices to users) were to drop (see figure 2).[188] The team explored three threshold launch prices to LEO: $1000/lb (believed to have a high probability of being achievable and a price that might double the mass in orbit over current levels); $600/lb (where new markets such as space business parks and hazardous nuclear waste disposal were estimated to be created); and $400/lb (where both new and old markets might show extensive growth).

But all three of the launch prices the CSTS investigated are significantly below most of the cur-

[187] The proposals generally call for initial deployment in clusters on MLVs or HLVs. Individual replacement satellites might be launched on SLVs. The LEO satellites are expected to have a life span of four to five years as opposed to 10 years for GEO satellites. Satellite replacement will require rapid launch responsiveness. Since there are a number of competing systems, launch costs will need to be minimized. See Ralph DePalma, U.S. Department of Transportation, "Responding to the Market," *U.S. Commercial Space Launch Industry*, 1995.

[188] Cost and price are often confused in the discussion of space transportation systems. Providers look at cost, while users focus on the price of launches. Even if launch costs were to be lowered dramatically, if launch prices were only lowered enough to capture much of the available market, the expansion of new markets might be very small.

TABLE 16: Commercial Satellites Launched by the United States and Arianespace

	U.S. commercial launches			Arianespace commercial launches			
Year	**European satellites**	**Other satellites**	**Total**	**European satellites**	**Other satellites**	**Total**	**Grand total**
1977	6	6	12	0	0	0	12
1978	3	6	9	0	0	0	9
1979	1	2	3	0	0	0	3
1980	0	2	2	2	0	2	4
1981	1	5	6	2	1	3	9
1982	0	10	10	2	0	2	12
1983	1	9	10	2	1	3	13
1984	2	9	11	4	2	6	17
1985	0	14	14	3	4	7	21
1986	0	1	1	3	2	5	6
1987	0	1	1	2	1	3	4
1988	0	0	0	6	7	13	13
1989	1	1	2	6	4	10	12
1990	4	7	11	7	9	16	27
1991	3	3	6	10	5	15	21
1992	3	5	8	5	8	13	21
1993	1	4	5	6	11	17	22
1994[a]	0	10	10	3	11	14	24

[a] Scheduled.

SOURCE: U.S. General Accounting Office, "National Space Issues: Observations on Defense Space Programs and Activities," GAO/NSIAD-94-253, August 1994, p. 26.

TABLE 17: Selected LEO Communications Systems

Type	System	Organization	Number of satellites	Orbital location	Projected initial operations
"Mega" LEO	Teledesic	Teledesic Corp.	840	LEO	1999
Big LEO	Iridium[a]	Iridium	66	LEO	1996
	Globalstar[a]	Loral/Qualcomm	48	LEO	1997
	Odyssey[a]	TRW	12	Medium orbit	1997
	Ellipso	Ellipsat Corp.	14-18	Elliptical	1996
Little LEO	Orbcomm	Orbital Communications	24	LEO	1994
	Starsys	NACLS, Inc.	24	LEO	1996
	Aries	Constellation Communications	48	LEO	1995

[a] Licensed by the FCC in January 1995.

SOURCES: U.S. Congress, Office of Technology Assessment, *The 1992 World Administrative Radio Conference: Technology and Policy Implications* (Washington, DC: U.S. Government Printing Office, 1993), p.114. Ralph DePalma, U.S. Department of Transportation, "Responding to the Market," *U.S. Commercial Space Launch Industry*, 1995. Frank C. Weaver, Director, Office of Commercial Space Transportation, U.S. Department of Transportation, "Dear Colleague," letter to industry addressing estimated LEO launch services, Mar. 13, 1994.

FIGURE 2: CSTS Medium Probability Market Distribution

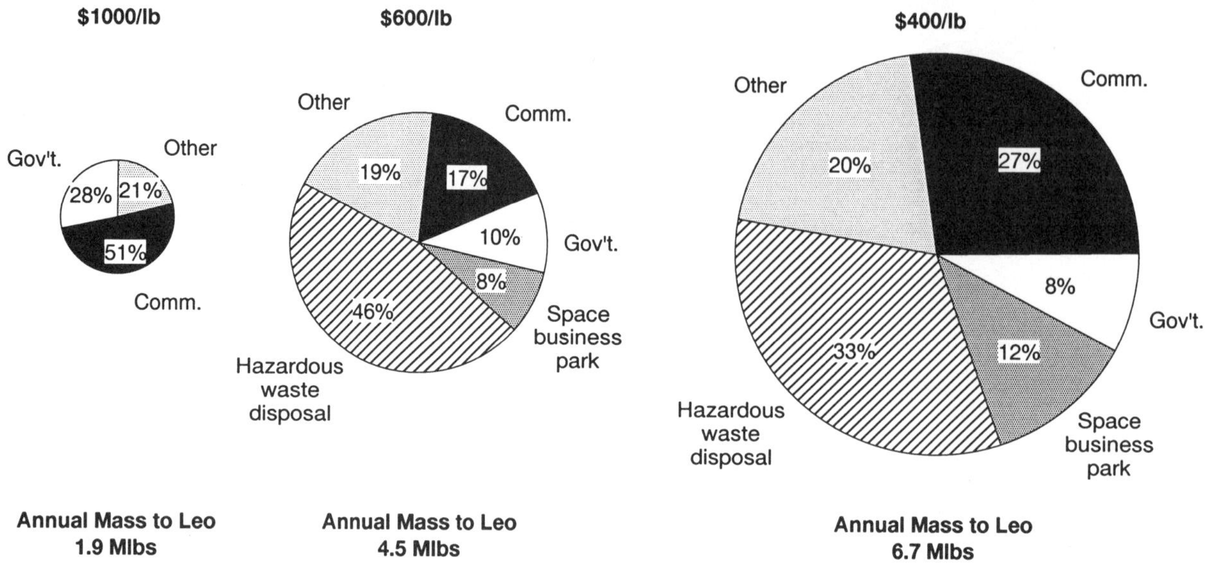

SOURCE: Commercial Space Transportation Study Alliance, "Commercial Space Transportation Study," Executive Summary, 1994, p. 8.

rent estimates for launch prices to LEO.[189] Thus, until a new vehicle can radically lower launch costs, the conservative market estimates appear appropriate and are supported by government findings. DOD's "Space Launch Modernization Study," for example, concluded that the commercial launch market has little potential for significant growth or economies of scale.[190] Long-term (15 to 20 years) growth, however, is an open question.

Factors other than price seem to influence the market share for U.S. firms. The CSTS identified several, including booking time, ability to launch on need, ability to launch on schedule, high launch rate, high reliability, simplified launch operations, and standardized payload interface (see table 18).[191]

A recent General Accounting Office (GAO) report[192] asked industry representatives why they had selected Arianespace over U.S. launchers. GAO reported that price was an important consideration, but respondents considered U.S. launchers competitive in price. Other factors considered were the aggressive and innovative marketing techniques of Arianespace, and the European firm's perceived space launch success rate.

---

189 U.S. Department of Defense, op. cit., footnote 15, p. II-10, cites prices to GTO of $12,000/lb for a U.S. provider, $8,000/lb as the Ariane goal, and $4,000/lb as the Chinese/Russian possibility. An analysis of several U.S. launch vehicles indicates that, in the best case, they might put roughly four times as much weight in LEO as in GTO. These estimates equate to roughly $3,000/lb, $2,000/lb, and $1,000/lb to LEO.

190 U.S. Department of Defense, op. cit., footnote 19, p. 9.

191 The Commercial Space Transportation Study Alliance, op. cit., footnote 16, p. 10.

192 U.S. General Accounting Office, "National Space Issues: Observations on Defense Space Programs and Activities," GAO/ NSIAD-94-253, Washington, DC, pp. 26-27. The GAO reported that the U.S. commercial launch industry is price competitive with foreign launchers and the future launch market appears limited. The report observed that "using requirements of the U.S. commercial launch industry as justification for developing a new launch vehicle does not appear warranted."

TABLE 18: Factors Employed in Launch Service Decisions

Factor	Source of difference	Potential influence on decision
Price	For firms basing price on market principles, different cost structures will result in different pricing. And every bid represents a unique situation. For firms operating on non-market principles, cost is often unknown and the price can be set indiscriminately.	Price is often the primary factor in the acquisition of space launch services. Price is affected by international exchange rates and the availability of competition.
Reliability	No one is quite certain of the actual reliability of launch vehicles. Most observers use launch success rate as a substitute indicator of actual reliability.	The expected value of a service varies with the probability of the outcomes. The higher the probability of success, the greater the value of the service.
Insurance rate	Varying launch success rates of launch vehicles (and other factors) lead the space insurance industry to charge different insurance rates for each launch vehicle.	The insurance rate can be thought of as a variable rate tax. It directly effects the perceived cost of the launch vehicle.
Interest rates	Interest rates vary with both time and place.	High interest rates increase overall launch costs, particularly for those with very long launch schedules.
Expected on-orbit life	More station-keeping fuel can be stored on a satellite being launched on a vehicle with higher lift capacity. Therefore, for a satellite of equivalent functionality, the on-orbit life expectancy of the spacecraft will vary with the launcher.	For a commercial satellite, the longer a satellite is in orbit, the greater its revenue stream (and hence profitability) will be. For civil and military payloads, longer on-orbit life translates into greater functionality or reduced program costs.
Launch schedule	Each launch vehicle adheres to a different launch schedule. Factors that affect launch schedule include manufacturing time, payload integration, launch licensing, and satellite licensing (in the event of export).	A long launch schedule may translate into lost functionality and/or revenues for a commercial satellite owner. Launch schedule is especially important when replacing a satellite rendered inoperable or approaching the end of its life expectancy.
Payment schedule	Each launch service provider structures a unique payment plan for its customers.	A payment schedule that pushes a good portion of the payments off into the future is advantageous to the customer because the customer can, in the meantime, invest the resources elsewhere.
Stand down time after a failure	Varying regulatory policies affect the probable period of time that a launch vehicle is made to stand down after a failure.	Longer than expected stand down times may result in the loss of vital functionality (commercial, civil, or military) and/or significant commercial revenues.
Responsiveness	The ability to offer a launch date specified by the customer.	Provider may be selected on the ability to launch a satellite in a certain quarter or year.
Technical	Launch vehicles have different sizes and potential payload configurations.	Satellite might be launched alone or bundled with other satellites. In the latter case, position in the stack may be important.

SOURCE: Office of Technology Assessment, 1995.

The examination of the potential commercial market raises a number of issues that might be considered in evaluating U.S. policy toward the private launch industry. One of the most important is to gain a better understanding of the inhibitors to U.S. business penetration of the global business. If price is a less important factor in customer decisionmaking than it is often alleged, then changes in design philosophy (e.g., that result in reductions in launch personnel) and accommodation of commercial needs (e.g., responsiveness) may be particularly important.

Overlap of Government and Commercial Markets

Any government-industry partnership is challenged by the apparent disparity between the missions and current payloads of the government, and the missions and payloads in the commercial market.

The "Space Launch Modernization Study's" description of the often conflicting goals and needs of the four major space launch user communities—defense, intelligence, civil, and commercial—is useful in considering the challenges in achieving synergy among space transportation systems. The report notes that the defense mission demands a launch capability that can lift a mixed collection of warning, surveillance, communications, weather, and navigation satellites into orbit in a timely manner. The intelligence community needs to lift relatively few payloads, which are typically large and very expensive. The nondefense (civil) government market has a special need for vehicles that can carry people back and forth to space. The commercial market is currently dominated by geosynchronous communications satellites, although the mix of satellites in the commercial market may change with the advent of new commercial LEO satellite communications constellations.

As noted above, DOD's overall EELV policy aims to bring down its overall space transportation budget and most of the savings are expected in the heavy payload range—not in the commercially compatible medium launchers.[193] This apparent mismatch of focus need not preclude some spinoff into the more commercially important sector—depending on the nature of the modifications made. DOD has generally been pursuing lighter satellites to meet future needs and thus use medium-lift capabilities, but intelligence satellites are still "generally large and expensive so that reliable, heavy-lift capability is a top concern."[194] Changing these configurations may be difficult and costly, or impossible. But the EELV family developed in the DOD program might be designed to take less time on the pad, require fewer personnel to launch, and be more economical to produce.

NASA's need for a vehicle that carries people complicates the connection with commercial space transportation service. Industry has little use for such a capability, but the development of a space tourism trade might pay for the marginal cost of outfitting vehicles to carry humans with reasonable safety.

Gaining the benefits of any synergy that might exist among the markets will demand a detailed investigation of the nature of the future market and the likely phasing of any expanded commercial market. A moderate expansion of the commercial market might occur with the development of a less costly family of EELV, but a greatly expanded market depends on reduced costs to orbit that are only thought possible with the development of a viable RLV or partially reusable launch vehicle. Moreover, these reduced costs have to translate into reduced prices for space transportation. A strategy of significantly reducing costs while only reducing prices to the point that a majority market share can be gained may not greatly expand the

[193] U.S. Department of Defense, op. cit., footnote 19, figure II-6, p. II-10. The bulk of the commercial market is in launching 3,000 to 10,000 lbs to GTO—a range covered by the Delta II, Atlas IIA, and Ariane 4.

[194] U.S. Department of Defense, op. cit., footnote 19, p.5. Moreover, these heavy launch vehicles could be made more compatible with commercial needs if they were used to launch two to three satellites at a time as is the case with Arianespace launchers.

overall use of space. Nonetheless, regardless of launch costs and the pricing policy followed, for the foreseeable future the government market will remain important to any private sector investment decision.

ISSUE 4b: *The government-industry relationship*

The NSTP and all the implementation plans stress the need for closer government-industry cooperation. Yet questions remain about how a future government-industry relationship can or should be structured. The NSTP designates the DOT and DOC to be the advocates of increased commercial participation, but with little money for space transportation and small staffs dedicated to space issues, DOT/DOC are unlikely to have much real impact. Indeed, although they participate in interagency discussions,[195] they are likely to find it difficult to influence relationships that NASA and DOD are already negotiating with their prime contractors.

NASA officials, at least at the highest levels, appear to support the idea that a strong, internationally competitive commercial space transportation base is essential not only for development of its future space transportation, but also to run the United States' future launch infrastructure. More important, NASA needs substantial private sector investment to build a new RLV to replace the aging Space Shuttle. The Agency is, therefore, seeking policies that will provide support for industry and provide incentives for industry to invest.

DOD, on the other hand, appears less concerned about developing a close partnership with industry than is NASA. To be sure, changes are occurring in DOD's relationship with industry as a result of general DOD acquisition reform, but these changes are more in line with developing a more efficient way of doing business than with organizing a development partnership. DOD, in contrast to NASA, does not have the same perceived need for a new space launch vehicle to perform its missions. Although its current fleet of ELVs are considered by the Department to be too costly and inflexible, they still perform well enough to meet the Department's fundamental mission requirements. The EELV program thus has a limited goal of allowing the Department to lower costs and gain some increased launch flexibility.

DOD's recent space launch vehicle industrial base assessment concluded that the U.S. space launch industry was reasonably profitable, had adequate production capacity, and was capable of meeting DOD's space launch requirements for the foreseeable future (see box 10).[196] The report concluded that:

> The U.S. space launch industry remains viable and capable of meeting DOD launch requirements. All three prime contractors currently supplying ELVs for DOD use are profitable, despite considerable production overcapacity in the large and small vehicle industry segments. Considerable industry consolidation is both inevitable and necessary. Overhead costs will be reduced and the Department, and ultimately the U.S. taxpayer, will benefit.[197]

Because current ELV production capacity can meet or exceed current demands (see table 19), the Department expects to benefit from the anticipated industry consolidation that will reduce overhead costs and prices.[198] Thus, while the Department generally supports enhanced commercial competitiveness in the space transportation industry, it sees no reason to provide specific financial support to achieve this goal.

---

[195] U.S. Department of Defense, op. cit., footnote 19, p. 6. DOT chairs the Interagency Coordination Committee on Transportation Research and Development.

[196] U.S. Department of Defense, op. cit., footnote 15, pp. ES-8-ES-10.

[197] Ibid., p. ES-13.

[198] Ibid., p. ES-10.

BOX 10: DOD's Assessment of the U.S. Space Transportation Industry

DOD recently conducted an assessment of the industrial base that supports space launch vehicles. The *Industrial Assessment for Space Launch Vehicles* is one of several defense technology and industrial base studies designed to provide DOD and Congress better insight into the changes that are occurring in the defense industry. The studies seek to identify essential capabilities that might be threatened with loss as a result of industrial restructuring and provide information for budget and program decisions to preserve needed capabilities. DOD's space launch assessment concentrated on domestic capabilities, the portion of the industry supporting ELV, prime contractors, and upper-tier subcontractors. The assessment also touched on the Space Shuttle base, space transportation infrastructure, and foreign sources of goods and services.

The assessment concluded that the industrial base supporting DOD's space launch needs was adequate to meet the Department's requirements for the foreseeable future. There is sufficient production capacity to meet DOD's expected demand for launch vehicles. Indeed, there is overcapacity in SLV and HLV, as shown in table 19.

DOD also reported that the four major ELV prime contractors (McDonnell Douglas, Lockheed, Martin Marietta, and Orbital Sciences)[1] were all profitable. Under the circumstances, DOD plans to let the base consolidate and allow the prime and first-tier contractors to ensure the availability of the subtiers since they "have all demonstrated an ability to manage the risks associated with a changing vendor base."[2]

[1] Lockheed and Martin Marietta have since merged into one company called Lockheed Martin.

[2] U.S. Department of Defense, Office of the Deputy Assistant Secretary of Defense (Industrial Affairs), "Industrial Assessment for Space Launch Vehicles," Washington, DC, January 1995, p. ES-10.

SOURCE: Office of Technology Assessment, 1995.

Industry sources argue, however, that the DOD's assessment of the health and profitability of the industrial sector is based on periods in which government business dominated and government investment and working capital were available. In the future, profit margins will have to be considerably higher for industry to attract substantial investment capital.

While the potential for moderately lowering launch costs (enough to achieve a positive return on investment in a reasonable period of time) and improving access to space might be sufficient to prompt government expenditures on new launch capabilities (if funds were available), it is not clear what levels of launch price reductions might have to be achieved to convince industry that significant commercial market expansion is likely and thus entice significant industry investments. An anticipated threefold reduction in cost has prompted a $100-million, private sector investment in the X-34. Commercial firms must evaluate their space transportation investment alternatives not only against alternative launch systems, but also against alternative investment opportunities outside the space industry.[199]

[199] Anthony L. Velocci, Jr., "Augustine Identifies Key Operating Themes," *Aviation Week and Space Technology*, Feb. 20, 1995, pp. 44-46. Norman R. Augustine, Chairman and CEO of Martin Marietta Corp. states the issue of investment alternatives quite clearly: "All of the U.S. companies that are in the launch vehicle business today have other alternatives other than launch vehicles. At Martin Marietta when we have a dollar to invest, we can invest it in our crushed rock business or in the business of developing more reliable, more efficient launch vehicles. We make those decisions in a fairly analytical business-like fashion. If at any point it looks like the launch vehicle business isn't viable, then we will invest our money in our highly profitable crushed rock operation."

TABLE 19: U.S. ELV Production Capacities

Launch vehicle	Launch vehicle class	Annual launch capacity (CCAFS + VAFB)[a]	Annual production capacity[b]	Annual U.S. launches[c]	Production over capacity[d]	Excess capacity factor[e]
Pegasus	Small	12	12-50	6	6+	2x
Taurus	Small	3	24	3	21	8x
Delta	Medium	12 + 6	12	11	1	Small
Atlas	Medium	10 + 4	8	8	0	None
Titan IV	Large	3-4 + 3-4	10	3	7	3x

[a] The maximum number of launches possible, given current facilities and personnel at both Cape Canaveral Air Force Station and Vandenberg Air Force Base, based on IDA analysis.

[b] Based on IDA analysis of contractor data. Does not include surge capability.

[c] Typical annual U.S. launches, 1995-2010. Apportionment of launches to specific vehicles within a launch vehicle class may vary. Includes firm, probable, potential, and launch-on-need launches. Launches in an individual year may vary.

[d] "Annual Production Capacity" minus "Annual U.S. Launches."

[e] "Annual Production Capacity" divided by "Annual U.S. Launches."

SOURCE: U.S. Department of Defense, Office of the Deputy Assistant Secretary of Defense (Industrial Affairs), "Industrial Assessment for Space Launch Vehicles," Washington, DC, January 1995, p. ES-10.

Precisely what return on investment a firm might need in order to conclude that a space transportation investment is appropriate is uncertain. Estimates discussed with industry during OTA's assessment ranged from 20 to 50 percent, depending on the perceived nature of the market risk. What is clear is that the private sector will need to be persuaded that there is reasonable potential for making a profit before firms will make significant investments. In the absence of the potential for clearly defined financial returns to industry, a true government-industry partnership will probably be elusive and remain a customer-provider relationship.

ORBITAL SCIENCES

The X-34 could ensure the United States' continued leadership in the SLV market.

In an attempt to develop this partnership, NASA reports that it has significantly modified its management structure for the RLV program. The Agency has, for example, centralized the program management under a small team at its headquarters. Yet critics argue that the current plan signals a less than radical departure from past NASA R&D programs, with technologies identified for research, and designated decision points.

Still, observers note that NASA is able to establish this fairly detailed research plan because of the experience gained earlier during NASP research. Further, these observers argue that NASA is now using the accumulated knowledge based on past research in an innovative (for NASA) approach involving greater industry participation Indeed, NASA has incorporated some changes in its X-33 and X-34 CANs, allowing for-profit firms to use independent research and development (IR&D) money as part of their corporate contributions. The Agency has also announced that it will apply the approach that DOD used in the DC-X to its management of the RLV program.

MCDONNELL DOUGLAS

During development, the DC-X program office consisted of only 10-12 managers.

NASA, according to these observers, is doing what it does best—funding research efforts in areas of particular relevance to a national need (in the case of the DC-XA, propulsion, vehicle structures, and operations technologies).

Others, however, argue that there is insufficient change and that continuing to run a research program in the established way will preclude the introduction of innovative technologies.[200] NASA's government-industry teams have been criticized as potentially diffusing responsibility to such a degree that no one can be held responsible for development decisions. These teams raise questions about technical data rights and the ability to diffuse technology that is developed. Finally, some argue that NASA may have too many programs (including the X-33, the X-34, the DC-XA, the Med-Lite ELV, and Shuttle upgrades), for the money available. Some industry observers argue that current development programs are inadequately funded and will never lead to production vehicles.

Further, the CSTS notes that launch infrastructure, principal launch assets, and manufacturing facilities are all currently under U.S. government control. Any new government-industry relationship, therefore, demands fundamental changes in ways of designing and operating space launch vehicles beyond simply modifying program management.

The achievement of large reductions in launch prices, however, might by itself radically change the government-industry relationship. The reductions could increase commercial business and result in significant industry independence from the government. For example, although advances in commercial electronics have made more electronics technology available to defense, it has also reduced DOD's leverage in guiding the direction of that industry's R&D and product development efforts. Under similar circumstances, commercial launcher design and operation might drift away from currently defined defense requirements. Unless government planners are able and willing to modify their requirements in concert with these trends, this reduced leverage might result in greater difficulty in meeting the government's goal of maintaining access to space.

---

[200] Some in industry complain that NASA will simply reject proposals as "nonresponsive" if they do not conform to NASA's preconceived concepts.

ISSUE 4c: *Risk management—striking the proper balance*

Many in industry argue that an anchor tenancy arrangement is essential to entice private sector investment in space transportation. By providing a guaranteed market for a specific period, anchor tenancy would reduce investment risk for the private sector while commercial markets are being established. Government purchases of semiconductors and computers played an important role in the development of these U.S. industries.[201] In some of those cases, however, both development funding and initial large purchases came from the government. In general, the U.S. government's track record in improving manufacturing competitiveness—as either the first customer for a new product or an important customer for established products—is viewed as weak.[202]

As noted earlier, anchor tenancy is often used to finance construction of buildings in the private sector and is sometimes used by the General Services Administration (GSA) for buildings that the government will subsequently occupy under lease. Legislation allows the GSA to enter long-term (up to 20 years) lease arrangements and to score the lease payments against its budget each year.

In the commercial world, aircraft firms usually also look for firm orders before making major investments in new aircraft models. This has recently occurred in commercial space launch, too, when McDonnell Douglas announced it will develop a new rocket (Delta III) with its own funds, but based on a firm commitment by the Hughes Telecommunications and Space Co. for 10 launches. Steven Dorfman, President of Hughes Telecommunications and Space noted that, "[b]y being the anchor customer for the Delta III we encourage McDonnell Douglas to make the investment to upgrade its highly reliable Delta to the Delta III class."[203]

But an anchor tenancy arrangement may be contentious. For example, both congressional and executive branch agencies with responsibility for examining the budget process have criticized the GSA leasing arrangements. They argue that such leases cost the government more than outright purchases because the private sector's cost of capital is always higher than the government's cost of capital. Moreover, critics[204] point out that showing outlays over time fails to capture the fiscal effect of the government's commitment, which occurs in the first year, not over the entire lease period.

In accordance with this view, many believe that anchor tenancy for space transportation would have to be considered a "lease-purchase" arrangement under the existing rules developed to record the effects on the budget of enacted and pending legislation ("scorekeeping rules"). Such arrangements are recorded in the budget as if the government had purchased the asset outright. The discounted present value of the expected costs of space launch services would be recorded as budget authority when the contract was signed. Outlays would be scored in proportion to construction activity on the launchers as if the government were

---

201 The government used early commercial computers for the Veteran's Life Insurance Program. This not only provided a market for the hardware, but also demonstrated a use that was subsequently adopted by commercial industry. U.S. Department of Commerce, personal communication, April 1995.

202 U.S. Congress, Office of Technology Assessment, op. cit., footnote 169, p. 70.

203 Hughes Space and Communication Co., "Hughes Buys 10 Launches as First Delta Customer," press release, Los Angeles, CA, May 10, 1995.

204 See James L. Blum, Deputy Director, Congressional Budget Office, "CBO Testimony on the Lease-Purchase Scorekeeping Rule," testimony presented at hearings before the Subcommittee on Legislation and National Security, Committee on Government Operations, House of Representatives, U.S. Congress, Washington, DC, Sept. 20, 1994; and Alice Rivlin, Acting Director, Office of Management and Budget, "Testimony on Hearing on H.R. 2680," testimony presented at hearings before the Subcommittee on Legislation and National Security, Committee on Government Operations, House of Representatives, U.S. Congress, Sept. 20, 1994.

building the system. A forthcoming Congressional Budget Office report examines this financing issue in detail.[205] This approach would make the use of anchor tenancy problematic if the sole objective of seeking private financing was to deal with current fiscal constraints in the NASA budget.

Other observers, however, argue that there is a need for new thinking on the anchor tenancy concept as it applies to high-technology systems. They note that this is particularly true in a period in which the government is slated to be less involved in the development of goods and services that the private sector might reasonably be expected to provide. Furthermore, they view GSA building lease practices as poor examples for what might work in space transportation, and argue that the development of new space transportation systems might be used to test alternative anchor tenancy concepts.

Advocates suggest several criteria for judging the appropriateness of anchor tenancy. They argue that anchor tenancy is inappropriate if there is significant technological risk. It should also not be used if it is simply a way to make a commitment outside the budget (concern over this is the reason for viewing anchor tenancy as a lease purchase). But, they believe that anchor tenancy might be successfully used if the situation is one in which there is little technological risk, the contractor is taking the risk of performance, the contractor is financing the project, and the contractor has design control. This situation would require competitive bidding to help determine what the financial markets believe to be an acceptable risk.

The competitive aspects might be handled in a number of ways. One observer has suggested developing an Access to Space Service Market (ATSSM) "open to commercial and government customers, and anchored by U.S. government demand."[206] The ATSSM would coordinate the needs of space launch users with current and future capabilities of space launch providers. Space launch contracts might be sold as firm contracts or convertible options promoting the development of a robust market for space transportation services. To develop sufficient demand, "the ATSSM must not be tailored for highly specialized demand such as high security or manned payloads."[207] In the end, however, it is possible that the investor's view of the current space transportation market might make it impossible to use anchor tenancy to develop a vehicle that would meet both unique government requirements and commercial needs.

Whatever type of anchor tenancy arrangement might be fashioned, it may be difficult to work out the details rapidly enough to support the government's current plan for RLV development. Industry analysts argue that the agreement will need to be executed before April 1996, when industry must begin to commit significant funds toward the development of the X-33 technology demonstrator.

Advocates argue that termination liability is essential for reducing the risk to the private sector of entering into a long-term agreement with the government. They cite the importance of risk reduction in drawing private sector investment. Skeptics, however, have argued that such arrangements amount to providing a "risk-free" environment for U.S. business. Still, termination liability is often a part of commercial and government contracts, but the liability usually only provides for money already spent, not for loss of future revenue, nor the cost of financing. Thus firms continue to carry some risk even if compensation for some, or all, of the funds already spent is guaranteed.

Using a large government prize to attract private sector funding of development is one of the most controversial ideas for financing a future RLV. Proponents argue that a prize would be a

205 U.S. Congress, Congressional Budget Office, *The Budgetary Treatment of an Option To Finance the Development of a New Launch System*, forthcoming, May 1995.

206 Charles W. Polk, "Buying Access To Space Rather Than Procuring a Space Transportation System," information paper, May 15, 1995.

207 Ibid.

great stimulus to investment and innovation, with many potential commercial spinoffs. Further, they argue that no government money would actually be spent until success is achieved.[208] Critics, however, argue that the uncertain market for space transportation, combined with the risk of losing such a contest, would scare away potential investors. And even if the contest proceeded, there might not be a winning result, or at least not soon enough to replace the Space Shuttle. NASA would have much less insight into the progress of the contestants, and its ability to meet civil space goals might be compromised should all fail to meet the stated objective. A further complication is determining the size of the prize that might entice bidders to the program without seeming extravagant to taxpayers.[209]

A space launch corporation is viewed by many as a potentially good management tool. This type of organization could deal directly with space transportation users and operate on business principles. The Moorman report, however, argued that it was unnecessary at this time unless there is a major breakthrough in the commercialization of space.[210] Further, DOD officials have expressed concern that a commercially focused corporation would concentrate on developments in more commercially useful MLVs and ignore the HLVs that are currently DOD's more costly problem.

Finally, some firms will make space transportation investments without any government guarantees. McDonnell Douglas' investment in the Delta III is one example. Boeing Corp. has announced it is joining with a European shipbuilder and two aerospace firms from the FSU to form SEA Launch—to launch satellites from a pad in international waters in the Pacific Ocean. And Kistler Aerospace has reported that it is planning to build a reusable rocket without government funds.[211] Such investments could result in further investments by competitors. Daniel Tellep, Chairman and CEO of Lockheed Martin Corp., has stated that the Delta III will cause Lockheed Martin to consider responses in its Atlas program.[212] If such commercial activities succeed, they could well lead to radical changes in government-industry relations.

ISSUE 4d: *Launch operations and infrastructure*

Many analysts argue that significant launch cost savings might be realized through changes in launch operations and infrastructure.[213] Efficient launch operations are key competitive advantages for both Ariane and Russia, and launch infrastructure design is an additional positive factor for Ariane. But important launch cost reductions[214] are unlikely unless launch operation engineers and fa-

---

[208] Although the money might not be spent, it would have to be authorized by Congress and become a budget item, possibly appropriated into an escrow account. A variant that is discussed uses anchor tenancy as the prize.

[209] Some proponents of government prizes have suggested a range of about $10 billion to $12 billion for a cheap, reusable demonstrator. They believe that a large prize would bring new players into the space transportation business.

[210] U.S. Department of Defense, op. cit., footnote 19, p. 6.

[211] Warren Ferster, "NASA Picks Three Teams for Phase 1 Design," *Space News*, Mar. 13, 1995, pp. 4, 36.

[212] Daniel Tellep, Chaiman, Lockheed Martin Corp., remarks at the Aerospace and Defense Financial Conference, Lionheart Research, New York, NY, May 10, 1995.

[213] See Bruce D. Berkowitz, "No Free Launch: Updating Space Infrastructure," *Issues in Science and Technology Policy* 10(2):76-81, winter 1993.

[214] Launch operations can be divided into several overlapping steps, including: processing and integration of the vehicle, processing and integration of the payload, launch management and control, post launch activities, and logistics. In the past, launch operations have been estimated to account for about 15 percent of Shuttle recurring costs and about 20 percent of Titan IV costs per flight. For an extended discussion of reducing launch costs through changes in launch operations, and infrastructure investments, see U.S. Congress, Office of Technology Assessment, *Reducing Launch Operations Costs: New Technologies and Practices*, OTA-TM-ISC-28 (Washington, DC: U.S. Government Printing Office, September 1988), pp. 3, 13.

ARIANESPACE

Modern, automated launch facilities in Kourou, French Guyana are a key factor in the commercial success of Arianespace.

cility managers have a greater role in the design of future launch systems.

In an attempt to capture some of these potential savings, the NSTP and the implementation plans address launch operations, infrastructure, and ground activities. All the current government space transportation programs include actions directed at improvements in these areas. The X-34 program, for example, includes vehicle health monitoring systems and ground operations/rapid-turn-around studies. The X-33 program includes launch operations as a key element. And the DOD implementation plan includes a description of over $1 billion in investments in infrastructure modernization and upgrades to be completed by FY 2004. Government seed money has supported the development of commercial spaceports. Modest government funding at the California Spaceport, for example, was followed by significant corporate investment by ITT and others.[215]

How future space transportation systems will operate, and how such operations will save money in comparison with current operations, might be key oversight issues for Congress. While government seed money appears to have been successful in drawing additional investment, the government may want to be careful not to create overcapacity through its use of funds. Facilities that are being developed will need to be designed to accommodate future space transportation needs.

Space transportation infrastructure includes a broader range of industries than simply those directly involved in the physical development, production, and servicing of vehicles. The availability of insurance at a reasonable and predictable cost is critical to commercial space ventures. Almost no company is willing to bet that its $200 million spacecraft will be safely placed into the intended orbit and will properly operate once there, without being protected by insurance. Insurance rates over the last decade have fluctuated between 6 and 30 percent, depending on the perception of risk and the availability of funding to support it. Maximum values capable of being insured have ranged from $100 million to $410 million, depending on market conditions. The cost and volatility of rates can be a barrier to the continued development of the U.S. commercial space industry.

Some of this volatility might be reduced if the pool of insured launches were expanded to allow government launches on commercial vehicles to

[215] The Spaceport, located at Vandenberg Air Force Base, CA, expects to be operational in 1996.

U.S. DEPARTMENT OF DEFENSE

Space Launch Complex 6 at Vandenberg Air Force Base in California is being converted into a commercial spaceport.

be underwritten by the insurance industry. The government and public would benefit from a quantified and limited cost associated with the risk of physical loss of hardware. The commercial space industry would benefit from a healthy and reliable space insurance market. This larger insurance pool could be developed through procurements that require performance requirements for the operation of a launch vehicle or a satellite in orbit. Insuring government payloads, however, would be an added cost to the government, at least in the near term.

ISSUE 4e: *Accommodating commercial needs*

The NASA and DOT/DOC implementation plans include extensive discussions of the partnership between government and industry to achieve future goals, yet there is no clear indication that the government is willing to compromise on its stated requirements in order to make use of more commercially responsive systems. Thus, one of the key questions is the extent to which its space transportation programs will be driven by rigid government space launch and payload requirements that provide minimal overlap with commercial space launch competitiveness issues.

For example, despite the rhetoric of government-industry partnership, NASA's program may be best structured to produce an RLV that will serve the U.S. government's space transportation needs first, rather than producing a commercially viable vehicle that will also meet government needs. One of the problems noted by industry is that NASA's development plans do not incorporate the means for industry to put payloads in GEO. To deal with this shortcoming, some industry officials have suggested developing, from the beginning, a medium-lift RLV that is capable of accommodating strap-on solid rocket motors and housing larger, heavier payloads in its payload bay. This option would allow designers to optimize the RLV for commercial use and still meet the government need to launch large and heavy payloads and service the International Space Station.

The recent NASA Med-Lite RFP is cited by some in industry as an example of an inability, or unwillingness, of the government to accommodate commercial operations. The RFP exceeded 325 pages. Industry argues that reducing government paperwork and unnecessary oversight is a major challenge that must be overcome if price competitive launch vehicles are to be developed.

The DOD EELV program is designed so that the "competition will encourage commercial innovation to expedite development and encourage cost saving."[216] But such savings are focused on development of a vehicle that will meet DOD requirements, not on a launch vehicle design optimized for commercial use. Further, while the DOD hopes to reduce space transportation costs significantly, these savings will probably be concentrated in the HLV range, where the majority of the DOD's funds are spent. While this concentration makes sense to DOD, it may do little to increase the ability of the U.S. space transportation business to compete internationally.

Still, DOD does report that it is applying advanced technology to reduce the size of its payloads and requirements for costly HLVs. It is also examining ways to reduce the costs associated

[216] U.S. Department of Defense, op., cit., footnote 126, p. 6.

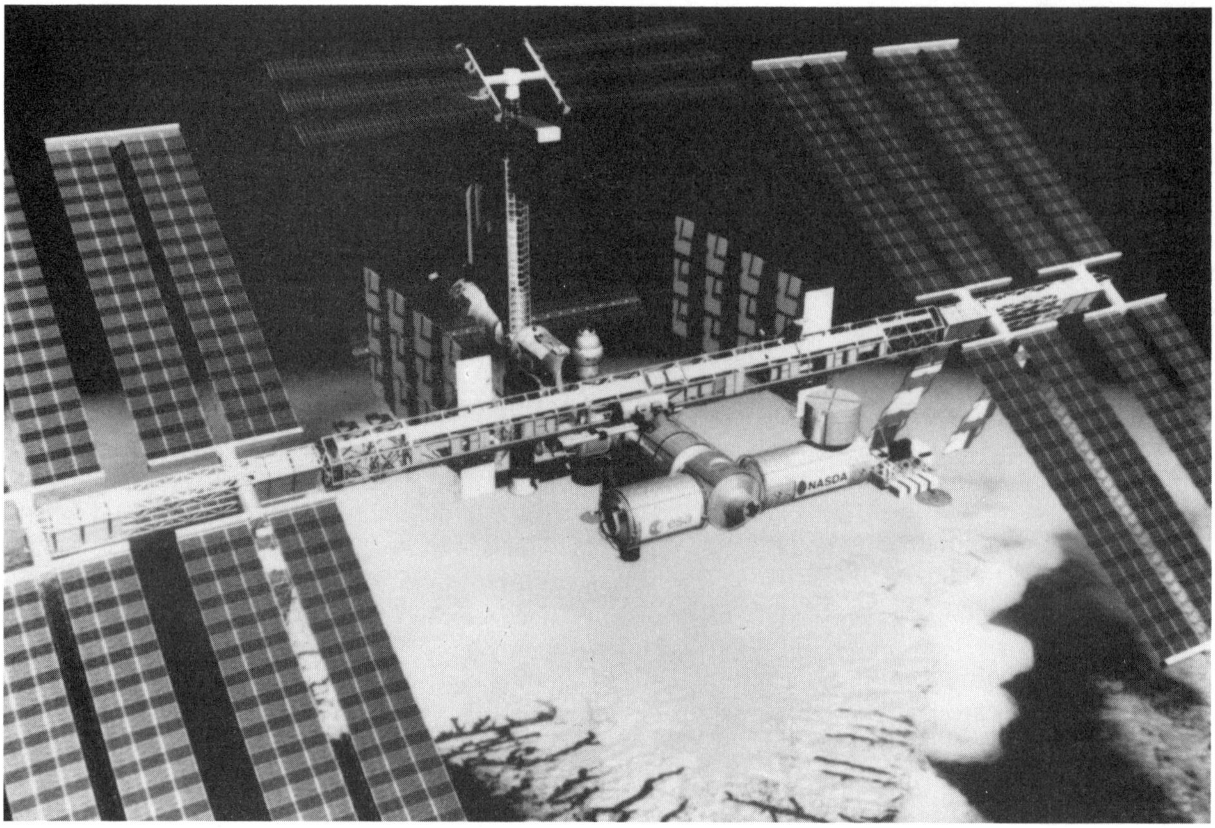

NATIONAL AERONAUTICS AND SPACE ADMINISTRATION

NASA plans on delivering crews and cargo to the International Space Station until, and possibly beyond, 2012.

with launching heavy payloads, which cannot be reduced to fit on SLVs or MLVs, so that it can leverage the competitive commercial market in order to reduce launch costs.[217]

As noted earlier, DOD has a detailed set of actions (that it views as useful to industry) to achieve its objectives. These include: a Titan IV program to ensure heavy launch capability; improving Atlas reliability; and upgrading Delta flight safety and avionics. ELV infrastructure is also being upgraded.

Additional Issues for Congress

As the previous sections discuss, the NSTP and its implementation plans cover a wide variety of issues of importance to the long-term health of the space transportation technology and industrial base. In this section, OTA identifies two issues of importance to Congress that were not addressed in either the policy itself, or the implementation plans.

ISSUE 5: *Preservation of long-range ballistic missile capabilities*

One critical component of the space transportation industry not addressed in either the policy or the implementation plans is the long-range ballistic missile segment of the space transportation technology and industrial base.[218] This omission may demonstrate just how far space launch ve-

[217] U.S. Department of Defense, op., cit., footnote 19, p. II-11.

[218] OTA intends to investigate the technological and industrial overlap between ballistic missiles and space transportation systems more fully in the main report of this assessment.

hicles have diverged from their ballistic missile roots, or it may be the result of a narrowly formed policy. In either case, failure to investigate possible connections between the development of long-range ballistic missiles and the development of launch vehicles risks overlooking policy options that could meet both government and commercial needs.

Originally, the military's requirement for long-range ballistic missiles created the space transportation technology and industrial base. The modern Atlas and Titan ELVs both evolved from early ICBMs, while the original Delta ELV used components from three ballistic missile.[219] Over time, these two segments of the technology and industrial base diverged, as ballistic missiles were optimized for round-the-clock readiness and quick launch which led to the development of solid rocket motors and required precision guidance. Launch vehicles, on the other hand, were modified to lift ever larger payloads into Earth orbit.

Still, considerable overlap between these two segments remains, particularly at the lower industrial tiers. For example, the new Lockheed Launch Vehicle (LLV) takes advantage of Trident II D5 technologies. Both the LLV and OSC's Taurus use the Thiokol Castor 120 solid rocket motor, which is derived from the Peacekeeper ICBM first stage.

As a result of nuclear arms control treaties and the end of the Cold War, the development and production of long-range ballistic missiles has almost halted. The Navy's Trident II D5 missile will continue to be produced into the next century at the rate of about 12 per year, with all production scheduled to stop in the year 2005. The Air Force is not currently producing any new missiles, but plans to modernize the Minuteman III by producing new motors and upgrading systems and materials. In addition, the Air Force plans to sustain the guidance and reentry vehicle industrial base.[220]

LOCKHEED MARTIN

Production of the Trident II D5 missile—the only long-range ballistic missile in production—is scheduled to stop in 2005.

The requirement to maintain the capability to design, test, and produce long-range ballistic missiles will continue as long as the United States depends on these missiles for part of its strategic deterrence. Eventually, the deployed missiles in the current forces will become obsolete and need to be replaced. Furthermore, new arms control treaties may compel the United States to deploy new single-warhead missiles, just as Russia is currently doing.

Both the Navy and the Air Force recognize the difficulty of maintaining production capability without producing anything. In the past, the Navy often had three ballistic missile programs running simultaneously. Both the Navy and the Air Force have instituted a series of programs to maintain what they consider the most critical elements of their missiles (e.g., guidance and reentry sys-

---

[219] Eugene M. Emme (ed.), *The History of Rocket Technology: Essays on Research, Development, and Utility* (Detroit: Wayne State University Press, 1964).

[220] William J. Perry, Secretary of Defense, "Annual Report to the President and the Congress," Washington, DC, February 1995, p. 90.

tems). OTA's past work on preserving industrial base capabilities, however, suggests that maintaining ballistic missile design and production teams may require new development and production opportunities.[221] Without these opportunities, scientists, engineers, and other workers may begin to seek challenges in other fields.

Under current designs, the U.S. long-range ballistic missile fleet depends on the production of solid rocket motors. A liquid-fueled EELV family and RLV follow-on to the X-33 could greatly reduce the demand for solid rocket motors, as could replacing the Shuttle's SRBs with liquid-fueled boosters. In that case, active production related to long-range ballistic missiles might be limited to the currently low-volume Taurus and LLV, as well as a few solid-rocket strap-on boosters, tactical missiles, or perhaps antiballistic missile interceptors.[222]

ISSUE 6: *The invisible lower industrial tiers*

Current policy and implementation plans appear to be principally directed at, and influenced by, the large prime contractors (e.g., Lockheed Martin, McDonnell Douglas, Rockwell International, and Orbital Sciences), yet there is far more to the industry than just those firms. The U.S. space transportation industry also includes the providers of components subsystems, such as Rocketdyne, Aerojet, and UTC-P&W for liquid-fueled engines; Thiokol, Hercules, UTC-CSD, and Aerojet for solid rocket motors; and hundreds of other lower-tier providers of goods and services. In fact, for each dollar spent on the procurement of space transportation services, roughly half flows down to second and lower tiers.[223]

An OTA workshop focused on activity in the space launch industry's lower tiers found great skepticism among many firms about the current government space transportation R&D programs. Participants from a wide spectrum of supporting industries reported that they: 1) doubt the government's commitment to build new space transportation systems (they believe the programs are too seriously underfunded to produce a vehicle), and 2) doubt that much of the R&D money will filter past the prime contractor level. Lower-tier firms that are highly dependent on space launch business are pessimistic about their survival. Those that are less dependent on the space transportation business plan to devote little effort to the current programs because few returns are anticipated.

Some representatives of the large, prime assemblers believe that the subcontractors will be there if the business is there. This may or may not be the case. If little or no money is available for the lower-tier firms in future development, those firms might shift their business elsewhere or cease to exist. Congress may wish to consider the possibility that key elements of the space transportation industrial base might be lost, just as other industrial sectors have lost important elements in the past (e.g., large diesel engines in the shipbuilding industry).

---

[221] U.S. Congress, Office of Technology Assessment, *Redesigning Defense: Planning the Transition to the Future U.S. Defense Industrial Base*, OTA-ISC-500 (Washington, DC: U.S. Government Printing Office, July 1991); U.S. Congress, Office of Technology Assessment, *Building Future Security: Strategies for Restructuring the Defense Technology and Industrial Base*, OTA-ISC-530 (Washington, DC: U.S. Government Printing Office, June 1992); and U.S. Congress, Office of Technology Assessment, *Assessing the Potential for Civil-Military Integration: Technologies, Processes, and Practices*, OTA-ISS-611 (Washington, DC: U.S. Government Printing Office, September 1994).

[222] OTA will examine both the preservation of long-range ballistic missile capabilities and the invisible lower tiers as part of its on-going assessment of the space transportation technology and industrial base.

[223] OTA analysis of the most recent (1987) unpublished U.S. Bureau of Economic Analysis input-output table.

Appendix: Clinton Administration Policy Statement and Fact Sheet

THE WHITE HOUSE

Office of Science and Technology Policy

For Immediate Release August 5, 1994

STATEMENT ON

NATIONAL SPACE TRANSPORTATION POLICY

The White House today released a new National Space Transportation Policy document, as developed by the National Science and Technology Council and approved by President Clinton. The policy sets a clear course for the nation's space program, providing a coherent strategy for supporting and strengthening U.S. space launch capability to meet the growing needs of the civilian, national security and commercial sectors.

The policy commits the nation to a two-track strategy of: (1) maintaining and improving the current fleet of expendable launch vehicles as necessary to meet civil, commercial, and national security requirements; and (2) investing R&D resources in developing and demonstrating next generation reusable space transportation systems with the potential to greatly reduce the cost of access to space.

The new policy accomplishes four fundamental objectives:

1) Establishes new national policy for federal space transportation spending, consistent with current budget constraints and the opportunities presented by emerging technologies. Under the new policy, DoD will assume the lead responsibility for modernization of the current expendable launch vehicle fleet. NASA will assume the lead responsibility for research and development of next generation reusable systems. NASA will focus their investments on technologies to support a decision no later than December 1996 on whether to proceed with a flight demonstration program. This program would, in turn, provide the basis for deciding by the end of the decade whether to proceed with a new launch system to replace the aging Shuttle fleet.

2) Establishes policy on federal agencies' use of foreign launch systems and components. With the end of the Cold War, it is important for the U.S. to be in a position to capitalize on foreign technologies — including Russian technologies — without, at the same time, becoming dependent on them. The policy allows the use of foreign components, technologies and (under certain conditions) foreign launch services, consistent with U.S. national security, foreign policy, and commercial space guidelines in the policy.

3) Establishes policy on federal agencies' use of excess U.S. ballistic missile assets for space launch, to prevent adverse impacts on the U.S. commercial space launch industry. Under START, these assets may be used in certain circumstances for civilian space launch. A serious concern in developing the policy was the possible impact that widespread government use of these assets could have on U.S. commercial launch companies. The policy obliges the government to fully consider commercial services as part of the decision making process and imposes specific criteria on the use of excess assets to avoid "flooding" the commercial market.

4) Provides for an expanded private sector role in the federal space transportation R&D decision making processes. In contrast with previous national policy on space transportation, this policy specifically directs the Departments of Transportation and Commerce to identify opportunities for government-industry cooperation and to factor these into NASA's and DoD's implementation plans.

These steps will help keep America at the forefront of space transportation technology, while ensuring that we have a robust and reliable expendable launch vehicle fleet.

Attachment:

National Space Transportation Policy fact sheet.

THE WHITE HOUSE

Office of Science and Technology Policy

For Immediate Release August 5, 1994

FACT SHEET

NATIONAL SPACE TRANSPORTATION POLICY

Introduction

The United States space program is critical to achieving U.S. national security, scientific, technical, commercial, and foreign policy goals. Assuring reliable and affordable access to space through U.S. space transportation capabilities is a fundamental goal of the U.S. space program. In support of this goal, the U.S. Government will:

(1) Balance efforts to sustain and modernize existing space transportation capabilities with the need to invest in the development of improved future capabilities;

(2) Maintain a strong space transportation capability and technology base, including launch systems, infrastructure, and support facilities, to meet the national needs for space transport of personnel and payloads;

(3) Promote the reduction in the cost of current space transportation systems while improving their reliability, operability, responsiveness, and safety;

(4) Foster technology development and demonstration to support future decisions on the development of next generation reusable space transportation systems that greatly reduce the cost of access to space;

(5) Encourage the cost-effective use of commercially provided U.S. products and services, to the fullest extent feasible, that meet mission requirements; and

(6) Foster the international competitiveness of the U.S. commercial space transportation industry, actively considering commercial needs and factoring them into decisions on improvements in launch facilities and launch vehicles.

This policy will be implemented within the overall resource and policy guidance provided by the President.

I. Implementation Guidelines

To ensure successful implementation of this policy, U.S. Government agencies will cooperate to take advantage of the unique capabilities and resources of each agency.

This policy shall be implemented as follows:

(1) The Department of Defense (DoD) will be the lead agency for improvement and evolution of the current U.S. expendable launch vehicle (ELV) fleet, including appropriate technology development.

(2) The National Aeronautics and Space Administration (NASA) will provide for the improvement of the Space Shuttle system, focusing on reliability, safety, and cost-effectiveness.

(3) The National Aeronautics and Space Administration will be the lead agency for technology development and demonstration for next generation reusable space transportation systems, such as the single-stage-to-orbit concept.

(4) The Departments of Transportation and Commerce will be responsible for identifying and promoting innovative types of arrangements between the U.S. Government and the private sector, as well as State and local governments, that may be used to implement applicable portions of this policy. U.S. Government agencies will consider, where appropriate, commitments to the private sector, such as anchor tenancy or termination liability, commensurate with the benefits of such arrangements.

(5) The Department of Defense and the National Aeronautics and Space Administration will plan for the transition between space programs and future launch systems in a manner that ensures continuity of mission capability and accommodates transition costs.

(6) The Department of Defense and the National Aeronautics and Space Administration will combine their expendable launch service requirements into single procurements when such procurements would result in cost savings or are otherwise advantageous to the Government. A Memorandum of Agreement will be developed by the Agencies to carry out this policy.

II. National Security Space Transportation Guidelines

(1) The Department of Defense will be the launch agent for the national security sector and will maintain the capability to evolve and operate those space transportation systems, infrastructure, and support activities necessary to meet national security requirements.

(2) The Department of Defense will be the lead agency for improvement and evolution of the current expendable launch vehicle fleet, including appropriate technology development. All significant ELV technology-related development associated with medium and heavy-lift ELVs will be accomplished through the DoD. In coordination with the DoD, NASA will continue to be responsible for implementing changes necessary to meet its mission-unique requirements.

(3) The objective of DoD's effort to improve and evolve current ELVs is to reduce costs while improving reliability, operability, responsiveness, and safety. Consistent with mission requirements, the DoD, in cooperation with the civil and commercial sector, should evolve satellite, payload, and launch vehicle designs to achieve the most cost-effective and affordable integrated satellite, payload, and launch vehicle combination.

(a) ELV improvements and evolution plans will be implemented in cooperation with the Intelligence Community, the National Aeronautics and Space Administration and the Departments of Transportation and Commerce, taking into account, as appropriate, the needs of the commercial space launch sector.

(b) The Department of Defense will maintain the Titan IV launch system until a replacement is available.

(4) The Department of Defense, in cooperation with NASA, may use the Space Shuttle to meet national security needs. Launch priority will be provided for national security missions as governed by appropriate NASA/DoD agreements. Launches necessary to preserve and protect human life in space shall have the highest priority except in times of national emergency.

(5) Protection of space transportation capabilities employed for national security purposes will be pursued commensurate with their planned use in crisis and conflict and the threat. Civil and commercial space transportation capabilities identified as critical to national security may be modified at the expense of the requesting agency or department. To the maximum extent possible, these systems, when modified, should retain their normal operational utility.

III. Civil Space Transportation Guidelines

(1) The National Aeronautics and Space Administration will conduct human space flight to exploit the unique capabilities and attributes of human access to space. NASA will continue to maintain the capability to operate the Space Shuttle fleet and associated facilities.

(a) The Space Shuttle will be used only for missions that requires human presence or other unique Shuttle capabilities, or where use of the Shuttle is determined to be important for national security, foreign policy or other compelling purposes.

(b) The National Aeronautics and Space Administration will maintain the Space Shuttle system until a replacement is available.

(c) As future development of a new reusable launch system is anticipated, procurement of additional Space Shuttle orbiters is not planned at this time.

(2) The National Aeronautics and Space Administration will be the lead agency for technology development and demonstration of next generation reusable space transportation systems.

(a) The objective of NASA's technology development and demonstration effort is to support government and private sector decisions by the end of this decade on development of an operational next-generation reusable launch system.

(b) Research shall be focused on technologies to support a decision no later than December 1996 to proceed with a sub-scale flight demonstration which would prove the concept of single-stage-to-orbit.

(c) Technology development and demonstration, including operational concepts, will be implemented in cooperation with related activities in the Department of Defense.

(d) It is envisioned that the private sector could have a significant role in managing the development and operation of a new reusable space transportation system. In anticipation of this role, NASA shall actively involve the private sector in planning and evaluating its launch technology activities.

IV. Commercial Space Transportation Guidelines

(1) The United States Government is committed to encouraging a viable commercial U.S. space transportation industry.

(a) The Departments of Transportation and Commerce will be responsible for identifying and promoting innovative types of arrangements between the U.S. Government and the private sector, as well as State and local governments, that may be used to implement applicable portions of this policy.

(b) The Department of Transportation will license, facilitate, and promote commercial launch operations as set forth in the Commercial Space Launch Act, as amended, and Executive Order 12465. The Department of Transportation will coordinate with the Department of Commerce where appropriate.

(c) U.S. Government agencies shall purchase commercially available U.S. space transportation products and services to the fullest extent feasible that meet mission requirements and shall not conduct activities with commercial application that preclude or deter commercial space activities, except for national security or public safety reasons.

(d) The U.S. Government will provide for the timely transfer to the private sector of unclassified Government-developed space transportation technologies in such a manner as to protect their commercial value.

(e) The U.S. Government will make all reasonable efforts to provide stable and predictable access to appropriate space transportation-related hardware, facilities, and services; these will be on a reimbursable basis. The U.S. Government reserves the right to use such facilities and services on a priority basis to meet national security and critical civil sector mission requirements.

(f) U.S. Government agencies shall work with the U.S. commercial space sector to promote the establishment of technical standards for commercial space products and services.

(2) U.S. Government agencies, in acquiring space launch-related capabilities, will, to the extent feasible and consistent with mission requirements:

(a) Involve the private sector in the design and development of space transportation capabilities and encourage private sector financing, as appropriate.

(b) Emphasize procurement strategies that are based on the use of commercial U.S. space transportation products and services.

(c) Provide for private sector retention of technical data rights, limited only to the extent necessary to meet government needs.

(d) Encourage private sector and State and local government investment and participation in the development and improvement of U.S. launch systems and infrastructure.

V. <u>Trade in Commercial Space Launch Service</u>

(1) A long term goal of the United States is to achieve free and fair trade. In pursuit of this goal, the U.S. Government will seek to negotiate and implement agreements with other nations that define principles of free and fair trade for commercial space launch services, limit certain government supports and unfair practices in the international market, and establish criteria regarding participation by space launch industries in countries in transition from a non-market to a market economy.

(a) International space launch trade agreements in which the U.S. is a party must allow for effective means of enforcement. The range of options available to the U.S. must be sufficient to deter and, if necessary, respond to non-compliance and provide effective relief to the U.S. commercial space launch industry. Agreements must not constrain the ability of the United States to take any action consistent with U.S. laws and regulations.

(b) International space launch trade agreements in which the U.S. is party must be in conformity with U.S. obligations under arms control agreements, U.S. nonproliferation policies, U.S. technology transfer policies, and U.S. policies regarding observance of the Guidelines and Annex of the Missile Technology Control Regime (MTCR).

VI. <u>Use of Foreign Launch Vehicles, Components, and Technologies</u>

(1) For the foreseeable future, the United States Government payloads will be launched on space launch vehicles manufactured in the United States, unless exempted by the President or his designated representative.

(a) This policy does not apply to use of foreign launch vehicles on a no-exchange-of-funds basis to support the following: flight of scientific instruments on foreign spacecraft, international scientific programs, or other cooperative government-to-government programs. Such use will be subject to interagency coordination procedures.

(2) The U.S. Government will seek to take advantage of foreign components or technologies in upgrading U.S. space transportation systems or developing next generation space transportation systems. Such activities will be consistent with U.S. nonproliferation, national security, and foreign policy goals and commitments as well as the commercial sector guidelines contained in this policy. They will also be conducted in a manner consistent with U.S. obligations under the MTCR and with due consideration given to dependence on foreign sources and national security.

VII. <u>Use of U.S. Excess Ballistic Missile Assets</u>

(1) U.S. excess ballistic missile assets that will be eliminated under the START agreements shall either be retained for government use or be destroyed. These assets may be used within the U.S. Government in accordance with established DoD procedures, for any purpose except to launch payloads into orbit. Requests from with the Department of Defense or from other U.S. Government agencies to use these assets for launching payloads into orbit will be considered by DoD on a case-by-case basis and require approval by the Secretary of Defense.

Mindful of the policy's guidance that U.S. Government agencies shall purchase commercially available U.S. Space transportation products and services to the fullest extent feasible, use of excess ballistic missile assets may be permitted for launching payloads into orbit when the following condition are met:

(a) The payload supports the sponsoring agency's mission.

(b) The use of excess ballistic missile assets is consistent with international obligations, including the MTCR guidelines and the START agreements.

(c) The sponsoring agency must certify the use of excess ballistic missile assets results in a cost savings to the U.S. Government relative to the use of available commercial launch services that would also meet mission requirements, including performance, schedule, and risk.

VIII. Implementing Actions

(1) Within 90 days of approval of this directive, United States Government agencies are directed to prepare the following for submission to the Assistant to the President for Science and Technology and the Assistant to the President for National Security Affairs:

(a) The Secretaries of Defense, Commerce, Transportation, and the Administrator of the National Aeronautics and Space Administration, with appropriate input from the Director of Central Intelligence, will provide a report that will include a common set of requirements and a coordinated technology plan that addresses the needs of the national security, civilian, and commercial space launch sectors.

(b) The Secretary of Defense, with the support of other agencies as required, will provide an implementation plan that includes schedule and funding for improvement and evolution of the current U.S. ELV fleet.

(c) The Administrator of the National Aeronautics and Space Administration, with the support of other agencies as required, will provide an implementation plan that includes schedule and funding for improvements of the Space Shuttle system and technology development and demonstration for next generation reusable space transportation systems.

(d) The Secretaries of Transportation and Commerce, with the support of other agencies as required and U.S. industry, will provide an implementation plan that will focus on measures to foster an internationally competitive U.S. launch capability. In addition, the Secretaries will provide recommendations to the Department of Defense and the National Aeronautics and Space Administration that promote the full involvement of the commercial sector in the NASA and DoD plans.

Superintendent of Documents **Publications** Order Form

Order Processing Code: P3

*7701

☐ **YES**, please send me the following:

Telephone orders (202) 512-1800
(The best time to call is between 8-9 a.m. EST.)
To fax your orders (202) 512-2250
Charge your order. It's Easy!

________ copies of ***The National Space Transportation Policy: Issues for Congress (128 pages),*** S/N 052-003-01415-7 at $9.00 each.

The total cost of my order is $________. International customers please add 25%. Prices include regular domestic postage and handling and are subject to change.

(Company or Personal Name) (Please type or print)

(Additional address/attention line)

(Street address)

(City, State, ZIP Code)

(Daytime phone including area code)

(Purchase Order No.)

Please Choose Method of Payment:

☐ Check Payable to the Superintendent of Documents

☐ GPO Deposit Account ☐☐☐☐☐☐☐ — ☐

☐ VISA or MasterCard Account

☐☐☐☐☐☐☐☐☐☐☐☐☐☐☐☐☐☐☐☐

☐☐☐☐ (Credit card expiration date)

Thank you for your order!

(Authorizing Signature) (5/95)

May we make your name/address available to other mailers? **YES** ☐ **NO** ☐

Mail To: New Orders, Superintendent of Documents, P.O. Box 371954, Pittsburgh, PA 15250-7954

THIS FORM MAY BE PHOTOCOPIED

Superintendent of Documents **Publications** Order Form

Order Processing Code: P3

*7701

☐ **YES**, please send me the following:

Telephone orders (202) 512-1800
(The best time to call is between 8-9 a.m. EST.)
To fax your orders (202) 512-2250
Charge your order. It's Easy!

________ copies of ***The National Space Transportation Policy: Issues for Congress (128 pages),*** S/N 052-003-01415-7 at $9.00 each.

The total cost of my order is $________. International customers please add 25%. Prices include regular domestic postage and handling and are subject to change.

(Company or Personal Name) (Please type or print)

(Additional address/attention line)

(Street address)

(City, State, ZIP Code)

(Daytime phone including area code)

(Purchase Order No.)

Please Choose Method of Payment:

☐ Check Payable to the Superintendent of Documents

☐ GPO Deposit Account ☐☐☐☐☐☐☐ — ☐

☐ VISA or MasterCard Account

☐☐☐☐☐☐☐☐☐☐☐☐☐☐☐☐☐☐☐☐

☐☐☐☐ (Credit card expiration date)

Thank you for your order!

(Authorizing Signature) (5/95)

May we make your name/address available to other mailers? **YES** ☐ **NO** ☐

Mail To: New Orders, Superintendent of Documents, P.O. Box 371954, Pittsburgh, PA 15250-7954

THIS FORM MAY BE PHOTOCOPIED